CONCEPTS OF E-COMMERCE

By

Adesh Kumar Pandey

Deptt. of Computer Science
(Krishna Institute of Engg. and Tech.)
Ghaziabad (U.P.)

S.K. KATARIA & SONS®

Publisher of Engineering Books

4885/109, Prakash Mahal, Dr. Subhash Bhargav Lane,
Opposite Delhi Medical Association, Daryaganj, New Delhi-110002 (INDIA)
Phone: +91-11-43551243, Mobile: +91-9871775858, 7678425209
e-mail: katariabook@gmail.com, katariabooks@yahoo.com
Website: www.skkatariaandsons.com
(e-books, subscriptions and bulk buying also available on our website)

Published by:

S.K. Kataria & Sons®

4885/109, Prakash Mahal, Dr. Subhash Bhargav Lane,
Opposite Delhi Medical Association, Daryaganj, New Delhi-110002 (INDIA)
Phone: +91-11-43551243, *Mobile:* +91-9871775858, 7678425209
e-mail: katariabook@gmail.com, katariabooks@yahoo.com
Website: www.skkatariaandsons.com

First Edition: 2007-2008; Second Edition: 2009-2010; Reprint: 2026

ISBN: 978-81-89757-81-6

Laser Typesetting by : AISA Enterprises, Delhi

Printed at : Al-Aziz Books, Sahibabad

Preface to the Second Edition

Now a days Electronic Commerce is critical factor to customer-to-business interactions, intra-business interactions, and business-to-business interactions. This book provides the basics to students about of E-commerce and it's implementation. The book is written to address undergraduate and post graduate students with a quick introduction to the major issues and topics in electronic commerce.

The book is written in very student friendly language and provides the complete basic knowledge of e-commerce. The book is organized and cover introduction to implementation of e-commerce.

Short question-answers and assignments are given in the book to review the complete concepts quickly. Suggestions are always welcome from the readers.

Adesh Kr. Pandey
AUTHOR

Acknowledgement

The book itself is an acknowledgement to the intensity, drive and technical competence of many individuals, who have contributed to it.

I am thankful to Almighty God, without whose grace nothing is possible for any one. I am very thankful to my friends and students who have inspired me to write this book.

I extend my gratitude to my wife Gunjan, my little daughter Shreya, my mother and my in-laws for their strength and support, failing which, this book wouldn't have been complete.

I convey my thanks to publisher of this book, for their constructive critism and suggestions for improvement in the second edition of the manuscript.

Adesh Kr. Pandey
AUTHOR

Contents

1. INTRODUCTION TO E-COMMERCE .... 1—11

1.1. History of Electronic Commerce 1
1.2. Electronic Commerce Definition 2
1.3. The Driving Forces for E-Commerce 2
1.3.1. Market and Economic Pressure 3
1.3.2. Societal and Environmental Pressures 3
1.3.3. Technological Pressures 3
1.3.4. Implications of Various Factors 4
1.4. Elements of E-Commerce 5
1.5. SWOT (Strength, Weakness, Opportunity, Trends) Analysis 5
1.5.1. Strength of E-Commerce 6
1.5.2. Weaknesses of E-Commerce 7
1.6. Opportunities of E-Commerce 9
1.6.1. Commercial Opportunities 9
1.6.2. Social Services 10
1.6.3. Entertainment 10
1.7. Trends In E-Commerce 10
■ **Summary** 11
■ **Review Questions** 11

2. E-COMMERCE FRAMEWORK .... 12—19

2.1. Electronic Commerce Framework 12
2.1.1. The Mechanism of Electronic Commerce 13
2.2. Types of Electronic Commerce 14
2.2.1. Business-to-Customer Electronic Commerce 14
2.2.2. Customer-to-Business Electronic Commerce 15
2.2.3. Consumer-to-Consumer Electronic Commerce 15
2.2.4. Business-to-Employee (B2E) Services 15
2.3. Intra/Inter Organizational E-Commerce 15
2.3.1. Use of E-Mail in E-Commerce 15
2.4. EDI (Electronic Data Interchange) 16
2.4.1. EDI Over Wan/Internet/Extranet 16
2.5. Electronic Funds Transfer (EFT) 17
2.6. Component of E-Commerce 18
2.6.1 Institutions of E-Commerce 18
■ **Summary** 19
■ **Review Questions** 19

3. ARCHITECTURAL FRAMEWORK OF E-COMMERCE .... 20—41

3.1. Internet 20
3.1.1. Internet History 20
3.1.2. Uniform Resource Locator (URL) 21
3.1.3. IP Addresses 22

3.1.4. Domain Names ... 22
3.1.5. Internet Infrastructure and Routing ... 23
3.1.6. Client/Server Model of Computing ... 23
3.2. World Wide Web ... 25
3.2.1. Web Architecture ... 25
3.2.2. The Web and E-Commerce ... 26
3.3. TCP/IP Protocols ... 27
3.4. Application Level Protocols ... 28
3.4.1. HTTP ... 28
3.4.2. Telnet Protocol ... 29
3.4.3. File Transfer Protocol ... 29
3.4.4. The SMTP Protocol ... 31
3.4.5. Secure Socket Layer Protocol ... 31
3.5. Web Browser ... 32
3.6. Web Servers ... 33
3.6.1. Role of Web Server on Internet ... 33
3.7. Common Gateway Interface ... 34
3.7.1. CGI Scripts, CGI Programs or CGI Applications ... 34
3.7.2. Requirements ... 35
3.8. Hypertext Markup Language (HTML) ... 35
3.9. Electronic Data Interchange ... 38
3.9.1. Building Blocks of EDI Systems : Layered Architecture ... 38
3.10. Transaction ... 40
■ **Summary** ... 41
■ **Review Questions** ... 41

4. NETWORK INFRASTRUCTUREFOR E-COMMERCE ... 42—52

4.1. Internet and Intranet Based E-Commerce Issues ... 42
4.2. Network for E-Commerce ... 44
4.2.1. Introduction to I-Way ... 44
4.2.2. Requirement of I-Way ... 44
4.2.3. Components of the I-Way ... 45
4.3. Introduction to Broadband ... 46
4.3.1. Working of Broadband ... 46
4.3.2. Advantages of Broadband ... 46
4.3.3. Types of Broadband ... 47
4.4. Introduction to ATM ... 49
4.5. Introduction to ISDN ... 49
4.5.1. Use of ISDN ... 49
4.5.2. Why ISDN to Access the Internet ... 50
4.6. Frame Relay Introduction ... 50
4.6.1. Frame Relay Standardization ... 51
4.6.2. Frame Relay Devices ... 51
■ **Summary** ... 52
■ **Review Questions** ... 52

5. MOBILE COMMERCE ... 53—66

5.1. Introduction to Mobile Commerce ... 53
5.2. M-Commerce Versus E-Commerce ... 54

5.3. Driving Force for M-Commerce 54
5.4. Limiting Factors 55
5.5. Mobile Customer and Applications 55
5.6. Applications of M-Commerce 56
5.7. Mobile Computing 56
5.7.1. Existing Cellular Network Architecture 57
5.7.2. Data Communications 58
5.7.3. CDPD Technology (Cellular Digital Packet Data) 59
5.7.4. Applications of Mobile Computing 61
5.7.5. The Future 62
5.8. Introduction to WAP (Wireless Application Protocol) 62
5.8.1. WAP Forum 63
5.8.2. Why do We Need WAP ? 63
5.8.3. WAP Architecture Overview 63
5.8.4. WAP Protocol Stack 64
5.8.5. Mobile Information Devices 65
■ **Summary** 66
■ **Review Questions** 66

6. WEB SECURITY ..... 67—77

6.1. E-Commerce Security Issue 67
6.1.1. Database Vulnerabilities 67
6.1.2. Shopping Cart Vulnerabilities 68
6.1.3. Customer's Passwords 68
6.2. Web Security 69
6.2.1. Types of Web Security Risks 69
6.2.2. Web Security Precautions 70
6.3. Client Server Technology 70
6.3.1. Overview of the Client Server Technology 71
6.3.2. Benefits 71
6.3.3. Concerns 72
6.3.4. Client Server Network Security 72
6.3.5. Emerging Client Server Security Threats 72
6.4. Firewalls and Network Security 74
6.4.1. Importance of Firewall 74
6.4.2. Firewall Protection 74
6.4.3. Limitations of Firewall 75
6.4.4. Types of Firewalls 76
■ **Summary** 77
■ **Review Questions** 77

7. ENCRYPTION ..... 78—95

7.1. World Wide Web Security 78
7.2. Internet Transactions and Security 78
7.2.1. Types of Identity Theft 79
7.2.2. Credit Card Fraud 80
7.2.3. How To Protect Ourself 81
7.3. Encryption 82
7.3.1. Elements of an Encryption System 82
7.3.2. Secret-Key Encryption 82
7.3.3. Public-Key Encryption 83

7.4. Digital Certificate 84
7.4.1. Use of Digital Certificates 85
7.4.2. Need of Digital Certificate 85
7.4.3. Digital Signature 85
7.5. Secure Socket Layer (SSL) 87
7.5.1. Use of SSL 88
7.6. Authentication 89
7.7. Introduction To VPN 89
7.7.1. Common Uses of VPNs 90
7.7.2. Basic VPN Requirements 91
7.7.3. Tunneling Basics 92
7.7.4. Security Mechanisms are Available Through VPN 93
7.7.5. VPN Management Issues 93
■ **Summary** 95
■ **Review Questions** 95

8. ELECTRONIC PAYMENT SYSTEMS ..... 96—117

8.1. Introduction 96
8.1.1. Problems with Traditional Payment Methods 97
8.2. Secure Electronic Transaction (SET) Protocol 97
8.2.1. A Sample SET Session 97
8.2.2. Objectives of SET 98
8.2.3. Mechanics of SET 98
8.2.4. An SET Implementation 99
8.2.5. Practical Requirements for SET 100
8.3. Payment Gateway 100
8.3.1. How Payment Gateways Work 101
8.3.2. Common Payment Gateways 101
8.4. Types of Electronic Payment System 102
8.4.1. Digital Token-based Electronic Payment Systems 103
8.4.2. Using the Digital Currency 103
8.5. Online Credit Card Based System 104
8.5.1. Types of Credit Card Payments 104
8.5.2. Technical Issues 105
8.5.3. Virtual PIN 106
8.5.4. DigiCash (or E-cash) 106
8.5.5. CyberCash/CyberCoin 107
8.5.6. PayPal 108
8.5.7. Smart Cards 108
8.5.8. Electronic Checks 108
8.6. Debit Cards Based EPS 109
8.7. Electronic Fund Transfer 110
8.7.1. EFT Services 110
8.7.2. Suggestions for EFT 110
8.8. Electronic Commerce and Banking 111
8.8.1. Changing Dynamics in the Banking Industry 111
8.8.2. Changing Consumer Needs 111
8.8.3. Cost Reduction 112
8.8.4. Home Banking Implementation Approaches 112
8.8.5. Management Issues in Online Banking 113

8.9. Electronic Commerce and Online Publishing 114
8.9.1. Need of Online Publishing 114
8.9.2. Online Publishing Approach 115
8.10. Electronic Commerce and Retailing 115
8.11. Online Shopping 116
■ **Summary** 117
■ **Review Questions** 117

9. NET COMMERCE ..... 118—128

9.1. Electronic Data Interchange 118
9.1.1. EDI Layered Architecture 119
9.1.2. EDI Versus E-mail 119
9.1.3. EDI in Action 120
9.1.4. Financial EDI 121
9.1.5. Standardization and EDI 122
9.2. Supply Chain Management 123
9.2.1. Basic Component of SCM 123
9.2.2. Supply Chain Management Software 123
9.2.3. Supply Chain Collaboration 125
9.2.4. Extended Supply Chain 125
9.2.5. Impact of Globalization on the Supply Chain 125
9.2.6. Some Emerging Technologies that will Affect the Supply Chain 126
9.3. Customer Relations Management (CRM) 126
9.3.1 People Aspect of CRM 126
9.3.2. Process and Technology Aspects to CRM 127
9.3.3. Issues of CRM 127
■ **Summary** 128
■ **Review Questions** 128

10. E-COMMERCE ISSUES, LAWS AND POLICIES ..... 129—143

10.1. Social Issues In E-Commerce 129
10.1.1. Definition of Ethics in E-Commerce 129
10.1.2. Major Threats and Challenges to Social Ethics 130
10.1.3. Improving Social Ethical Climate 130
10.1.4. Code of Social Ethics 131
10.2. Privacy Issue in E-Commerce 131
10.3. E-Commerce Laws 131
10.3.1. Legal Issues 132
10.3.2. Forms of Agreements 135
10.3.3. Encryption and Decryption Laws 137
10.3.4. United Nations on E-Commerce 138
10.3.5. International Chamber of Commerce (ICC) 138
10.3.6. E-Commerce Laws in India 138
10.4. E-Commerce: Government's Policies and Agenda 140
10.4.1. Role of Governments to Promote Electronic Commerce 140
10.4.2. Defining the Policy and Regulation for Electronic Commerce 140
10.4.3. Policy Recommendations for E-Commerce Growth 141
■ **Summary** 143
■ **Review Questions** 143

11. IMPLEMENTATION OF E-COMMERCE .... 144—151

11.1. Designing of a Well-Functioning Commercial WebSite 144
11.2. Developing E-Commerce Enabled Application 145
11.2.1. Getting an Internet Connection 146
11.2.2. Getting a Merchant Bank Account 147
11.2.3. Web Hosting 148
11.2.4. Registering a Domain Name 148
11.2.5. Obtaining a Digital Certificate 148
11.2.6. Finding a Provider of Online Transactions 149
11.2.7. Finding Shopping Cart Software 150
■ **Summary** 151
■ **Review Questions** 151

APPENDICES .... 153—170

EXAMINATION PAPERS .... 171—173

INDEX .... 174—178

CHAPTER 1

INTRODUCTION TO E-COMMERCE

INSIDE THIS CHAPTER

- 1.1. History of Electronic Commerce
- 1.2. Electronic Commerce Definition
- 1.3. The Driving Forces for E-Commerce
- 1.4. Elements of E-Commerce
- 1.5. SWOT (Strength, Weakness Opportunity, Trends) Analysis
- 1.6 Opportunities of E-Commerce
- 1.7. Trends In E-Commerce

1.1. HISTORY OF ELECTRONIC COMMERCE

We know that every invention has a perfect motive behind it , so as the E-Commerce. The need for electronic commerce stems from the demand within business and government to make better use of computing and to better apply computer technology to improve customer interaction, business processes and information exchange both within an enterprise and across enterprises.

Let us make some observation about the progress of E-commerce as follows :

During the 1970s, the introduction of electronic fund transfer (EFT) between banks over secure private networks changed financial markets. Electronic fund transfer optimizes electronic payment with electronically provided remittance information. Today there are many EFT variants, including the debit card whose use is becoming important at point of sales (POS) in grocery stores and retail outlets and direct deposits to employee bank accounts. Each day, every $4 trillion change hands via EFT over the computer networks linking banks, automated clearing houses and companies. The U.S. Treasury department estimates that 55 percent of all payments by the federal government in 1995 were made by EFT.

During the late 1970s and early 1980s, electronic commerce becomes widespread within companies in the form of electronic messaging technologies: electronic data interchange (EDI) and electronic mail. Electronic messaging technologies streamline business processes by reducing paperwork and increasing automation. Business exchanging traditionally conducted with paper such as checks, purchase orders and shipping documents are conducted electronically. Electronic data interchange allows companies to send/receive business documents (such as purchase orders) in a standardized electronic form to/from their suppliers.

In the late 1980s and early 1990s electronic messaging technologies become an integral part of workflow or collaborative computing systems (also called groupware). A prominent example of such systems is lotus notes. Groupware focused primarily on talking existing nonelectronic methods and grafting them onto an electronic platform for improved business process efficiency. Although hyped as the "killer app" in the early 1990s groupware efforts resulting in small gains in productivity and efficiency.

In the mid 1980s, a completely different type of electronic commerce technology spread among consumers in the form of online services that provided a new form of social interaction (such as chat rooms and inter-relay chat [IRC]) and knowledge sharing (such as new groups and file transfer programs). Social interaction created a sense of virtual community among the cyberspace inhabitants and helped give rise to the concept of a "global village".

In the 1990s, the advent of the worldwide web on the internet represents a turning point in electronic commerce by providing an easy-to-use technology solution to the problem of information publishing. The web made electronic commerce a cheaper way of doing business (economies of scale) and enabled more diverse business activities (economies of scope).

1.2. ELECTRONIC COMMERCE DEFINITION

IBM's Definition of E-Commerce is :

"The transformation of key business processes through the use of internet technologies"

According to R. Kalakota (1999) :

"E-Commerce is buying and selling activities over digital media"

E-Business – encompasses e-commerce and "includes the front and back-office applications that form the engine for modern business. Its not just about e-commerce transactions, it's about redefining old business models with the aid of technology to maximise customer value"

We can define electronic commerce in following three prospects :

1. From a **communication perspective**, electronic commerce is the delivery of information, products/services, or payment via telephone lines, computer networks or any other means.
2. From a **business process perspective**, electronic commerce is the application of technology toward the automation of business transactions and workflows.
3. From a **service perspective**, electronic commerce is the tool that addresses the desire of firms, consumers and management to cut service costs while improving the quality of goods and increasing the speed of service delivery.

Broadly speaking, electronic commerce emphasizes the generation and exploitation of new business opportunities and to use popular phrases, **"generate business value"** or **"do more with less"**.

These improvements may result in more effective performance **(better quality, greater customer satisfaction and better corporate decision making)**, greater economic efficiency **(lower costs)**, and more rapid exchange **(high speed, accelerated or real-time interaction)**. More specifically, electronic commerce enables the execution of information online transactions between two or more parties using inter-connected networks. E-commerce networks can be a combination of telephone system, cable TV, leased lines and wireless.

1.3. THE DRIVING FORCES FOR E-COMMERCE

Today's business environment is influenced greatly by market, economical, social and technological factors creating a lot of competition. Also these factors are unpredictable in that they may change unexpectedly at any time. Such changes in the business world have been defined as business pressure. In order to succeed in this dynamic world companies must take action such as lowering costs and improving quality etc. At the same time they need to be innovative in customizing their products, creating new products and providing good customer service.

The various environmental business pressures on companies today can be grouped into three categories : **Market, social and technological.**

These are discussed below :

1.3.1. Market and Economic Pressure

Now a days business is a pressure game. One must be ready to face utter competition in market. Now almost no company can sustain on the basis of monopoly since every product either manufactured of marketed by different companies. Let us see some parameters, which works as driving force for market :

1. **Strong Competition :** Now competition is the part of any business and companies are implementing new techniques every day to beat their competitors. E-Commerce is one of the tools of recent days, which is adopted by the companies.

2. **Global Economy :** In present scenario world becomes a global village and there are standard norms in market for business, so every country must follow theses norms. There is huge gap between the growth rate of developed and developing nations, so more and more developed nations are investing in every part of world. E-Commerce is the need of such a scenario.

3. **Extremely Low Labour Cost in Some Countries :** Now this the common feature of business that companies are having manufacturing units in one part of the world and the same product is marketed in other part of the world. Definitely some medium is required to maintain it and e-commerce is providing such medium.

4. **Frequent Changes in Market Demands :** As we discussed above this the age of technology. So any change which takes place in one part of the world, rapidly reflects every where and e-commerce is the tool which is necessary to sustain in present fast changing world.

5. **Increasing Expectations of Consumers :** The expectations of consumers about quality and services are very high, so it works like driving force for e-commerce.

6. **Awareness Among Consumers :** Now consumers are well aware. Internet, print media and electronic media keep update the consumers about new products and their rates.

1.3.2. Societal and Environmental Pressures

Society, Environment and Government are the part of business. These are the factors, which works like driving force for e-commerce. Let us discuss these in detail as follows :

1. **Government Regulations :** Every country have it's own rules and regulations but now a days almost every country have well defined cyber laws which provides a right environment for the e-commerce.

2. **Reductions in Government Subsidies :** Now world is very open, some countries provides lot of rebate in tax and in other forms to do business. But some other countries reducing the soaps for business. E-Commerce provides the opportunity to take theses challenges.

3. **Rapid Political Changes :** Whenever government changes in any country, there are some changes in guidelines and polices towards the business. It is also work as driving force for the e-commerce.

1.3.3. Technological Pressures

This is most critical factor for e-commerce. In present time every day new, cheap and more secure technology is coming in the market which makes e-commerce and cheap and secure medium of business.

1. Rapid Technological Changes : As we discussed above technology is the factor, which provides e-commerce an opportunity to become an perfect solution for business.

2. New Technologies : New and secure technology available in the market, which provides an easy, cheap and secure platform for the e-commerce.

3. Information Overload : By using e-commerce we can easily an efficiently manage information about the product and other business related things. This is a cheap and secure medium for managing important information.

4. Digital Convergence : Electronic commerce and the multimedia revolution are deriving the previously disparate industries such as communications, entertainment, publishing and computing worlds into ever-closer contact, forcing industries with traditionally different histories and cultures to compete and cooperate.

The emergence of multimedia standards and the shift to distributed computing and interworking are providing the raw power for the "digital convergence". Convergence has two dimensions: convergence of contents and convergence of transmission technology.

***(a) Convergence of Content*:** Regardless of its original form, convergence of contents ensures that digitized information can be processed, searched, sorted, enhanced, converted, compressed, encrypted, replicated and transmitted at low cost; this tool has profound implications for content-based industries like newspapers, magazines and books.

Convergence of contents enables sophisticated information publishing and browsing tools. Content convergence also enables companies to use networked databases and electronic publishing to improve corporate and individual decision-making and information processing. Content convergence facilitates the creation of a computer and network infrastructure that enables the coordination and integration of business processes or workflows.

***(b) Convergence of Transmission* :** Convergence of transmission compresses and stores digitized information so that it can travel through existing phone, wireless and cable wiring systems. Convergence of transmission is a convergence of communication equipment that provides the "pipeline" to transmit voice, data, image and video over the same line. Transmission convergence over a single line makes it easier to connect computers, high-speed peripherals and consumer electronic devices and to enhance a wide range of image-intensive or multimedia applications. Convergence of transmission results in easier access to networks and in the creation of new, low-cost delivery channels for new and old products aimed at either existing customers or new customer segments.

1.3.4. Implications of Various Factors

Economic and marketing forces and digital convergence have influenced the E-commerce. Now industries are repositioning themselves to take advantages of new opportunities, including the creation of entirely new services delivery channels, the development of new markets for existing products and the development of new information-based products for the online environment. For instance, digital convergence is reshaping the competitive environment for telecommunication services around the globe. In response to the intensified competition and reduced margins on basic telephone services in telecom markets, networks operators are building new computer-driven intelligent networks in order to offer a wide range of value-added services.

1.4. ELEMENTS OF E-COMMERCE

When we get down to the actual elements of commerce and commercial transactions, things get slightly more complicated because we have to deal with the details. However, these details boil down to a finite number of steps. The following list highlights all of the elements of a typical commerce activity. In this case, the activity is the sale of some product by a retailer to a customer :

1. If we would like to sell something to a customer, at the very core of the matter is the something itself. We must have a **product or service to offer.**
2. We must also have a **place** from which to sell your products. A place may be a phone number to put your order. A place may be store or shop for physical products.
3. We need to figure out a way to get people to come to our place. This process is known as **marketing**. If no one knows that your place exists, we will never sell anything. Locating your place in a busy shopping center is one way to get traffic. Sending out a mail order catalog is another. There is also advertising, word of mouth and even the guy in a chicken suit who stands by the road waving at passing cars.
4. We need a way to **accept orders**. At Wal-Mart this is handled by the check out line in a mail order company the orders come in by mail or phone and are processed by employees of the company.
5. We also need a way to *accept* **money**. If we are at Wal-Mart we know that we can use cash, check or credit cards to pay for products. Business-to-business transactions often use purchase orders. Many businesses do not require we to pay for the product or service at the time of delivery, and some products and services are delivered continuously (water, power, phone and pagers are like this). That gets into the whole area of **billing** and **collections.**
6. We need a way to deliver the product or service, often known as **fulfillment**. At a store like Wal-mart fulfillment is automatic. The customer picks up the item of desire, pays for it and walks out the door. In mail-order businesses the item is packaged and mailed. Large items must be loaded onto trucks or trains and shipped.
7. Sometimes customers do not like what they buy, so we need a way to **accept returns**. We may or may not charge certain fees for returns, and we may or may not require the customer to get authorization before returning anything.
8. Many products today are so complicated that they require **customer service** and **technical support** departments to help customers use them. Computers are a good example of this sort of product. On-going products like cell phone service may also require on going customer service because customers want to change the service they receive over time.

We find all of these elements in any traditional mail order company. Whether the company is selling books, consumer products, information in the form of reports and papers, or services, all of these elements come into play.

1.5. SWOT (STRENGTH, WEAKNESS, OPPORTUNITY, TRENDS) ANALYSIS

In this section we will fully analyze the E-commerce in the reference of following :

(1) Strength

(2) Weakness

(3) Opportunity

(4) Trends

The study of E-Commerce on theses parameters is called SWOT analysis.

1.5.1. Strength of E-Commerce

There are various attractions of e-commerce that has made it a popular Business tool. These are :

(i) **Economy :** E-commerce is highly economical. In e-commerce there is no rental of physical store space. Insurance or infrastructure investment. All we need is an idea, a unique product and a well-designed Web storefront to reach your cyber-customers, plus a partner to do fulfillment.

(ii) **Lower Cost :** Doing e-business on the Internet is extremely cost effective; it reduces logistical problems and puts a small business on a par with giants like Amazon.com, general Motors or Bank of America. In a commercial bank.

(iii) **Better Customer Services :** E-commerce emphasizes better and quicker customer service. Web-based customer service makes consumers happier. Instead of calling our company on the phone, holding for 10 minutes, then getting to a clerk who tapes into your account, the web merchant gives customers direct access to their personal accounts over the web. It saves time and money. It is a win-win proposition, for companies that do business with other companies, adding customer service to the web is a competitive advantage. The overnight package delivery service, where tracking numbers allow customer to check the whereabouts of a package online, is one good example.

(iv) **Greater Profit Margin :** E-commerce means greater profit margins. For example, the cost of processing a conventional airline ticket is Rs. 400. According to one travel agency, processing the same ticket (called e-ticket) over the web costs Rs. 50 only. Along with higher margins, business can gain more control and flexibility and are able to save time when manual transactions are done electronically.

(v) **Knowledge Markets :** E-commerce helps create knowledge markets. Small groups inside big firms can be funded with seed money to develop new ideas. For example, Daimler Chrysler has created small teams to look for new trends and products. A Silicon valley team is doing consumer research on electric cars and advising car designs.

(vi) **Swapping Goods and Services :** Swapping is trading something we have for something we want more. Ram's a networking consultant, offers his technical services through a company. People pay currency into Ram's account in exchange for his services instead of accepting the cash.

(vii) **Information, Sharing, Convenience and Control :** Electronic market place improve information sharing between merchants and customers and promote quick, just-in-time deliveries. Conveniences for the customer is a major driver for changes in various industries: Customers and merchants save money; are online 24 hours a day; 7 days a week; experience no traffic jams, no crowds and do not have to carry heavy shopping bags.

Control is another major driving factor. For example, instead of banks controlling the relationships with the customer, customers today can have more control of their banking needs via Internet Web sites. Banks like Bank of America and ICICI now give customers access to their accounts via the Web.

(viii) **Quick Comparison Shopping :** E-commerce helps customers to comparison between various options. Customers can visit for various option online, without moving from their places. They can compare rate and qualities through web only.

(ix) **Teamwork :** E-commerce helps people work together. E-mail is one example of how people collaborate to exchange information and work on solutions. It has transformed the way

organizations interact with suppliers, vendors, business partners and customers. More interaction means better overall results.

(*x*) **Productivity Gains :** E-commerce means productivity gains. Weaving the Web throughout an organization means improved productivity. Take the example of IBM, which incorporated the web into every corner of the firm-products, marketing and practices. The total cost savings in 1999 alone was close to $1 billion.

(*xi*) **Customization :** Digital products are highly customizable. They are easy to recognize, revise or edit. With information about consumer tastes and preferences, products can be differentiated (customized) and matched to individual needs.

(*xii*) **Ensure Secrecy :** EC devices invariably have in built security measures. For example, password, encoding , cryptography, cipher etc. are some of the mechanism/measures, which provide security and prevent unauthorized access and use of data, information and transactions.

Strength of E-Commerce Over Traditional Business can be more clear by the Compression of Both. Online purchasing and selling through e-business (wholesale and retail) Offers opportunities and advantages to the companies of all sizes engaged in different portfolios. Some of these advantages are enumerated below :

1. Sampling of products such as books, recorded lecturers, and music cassettes is possible on Web for business promotion.
2. Business houses selling their goods through catalogues, can reach additional global customers at lower cost.
3. For items subjected to frequent changes e.g. shares and bonds etc., the changes in rates/ quotations can be known instantly.
4. Customer's knowledge can be enhanced on topics such as news about local event, market research, industrial report, software etc., which can be distributed easily over the Internet.
5. A closer relationship can be developed amongst business sellers whose customer base is in on the Internet.
6. For wholesalers, distributors, retailers etc., it is convenient and efficient to buy from Web sites.
7. High quality, specialized products can be easily sold on Internet by the retailers.
8. Organizations can conduct interviews through Internet to select suitable candidate(s) for their requirements.
9. Group discussions can be conducted for corporate offices, industrial houses, business firms, universities, etc.

1.5.2. Weaknesses of E-Commerce

There are some problems and drawbacks of e-commerce and these are as follows :

(*i*) **High Risk of Internet Start Up Organizations :** Many stories unfolded in 1999, about successful executives in established firms leaving for Internet start-ups, only to find out that "get rich" dream with a dot.com was just that a dream. However, many dot.com organizations bubble busted in 2000 and onward due to various reasons like lack of good revenue model, everything is not possible through dot.com, problem related to customer satisfaction etc.

(*ii*) **Lack of a Blueprint for Handling E-Commerce :** There is a continuing shortage of *e-literate* people in the workplace. In a survey published in *Computer world*, nearly nine out of ten respondents said only of a few of their key managers have e-commerce skills, Internet experience, and foresight. Sixty-six percent also said that they are having a tough time attracting people wanting to take advantage of online opportunities. Finally, traditional organizational structures and cultures were found to inhibit progress in e-commerce.

(iii) **E-commerce is Not Free :** So far, success stories in e-commerce have favored large businesses with deep pockets and good funding. According to a recent report, small retailers that go head-to-head with e-commerce giants are fighting a losing battle.

(iv) **Security :** Security continues to be a problem for online business. In a 2000 *economist* article, 95 percent of Americans expressed reluctance to give out their credit card numbers via the Internet. For millions of potential cyber-customers, the fear of credit card theft is a real one. Consumers have to feel confident about the integrity of the process before they commit to the purchase.

(v) **Customers Relations Problems :** Not many businesses realizes that even an e-business cannot survive over the long term without loyal customers.

(vi) **System and Data Integrity :** Data protection and the integrity of the system that handles the data are serious concerns. Computer viruses are rampant, with new viruses discovered every day. Viruses cause unnecessary delays, file backups, storage problems and the like. The danger of hackers accessing files and corrupting accounts adds more stress to an already complex operation.

(vii) **Products People Won't Buy Online :** Imagine a Web site called furniture.com or living.com, where venture capitalists are investing millions in selling home furnishings online. In the case of a sofa, you'd want to sit on it, feel the texture of the fabric, etc. Beside the "sofa-road-test" factor, online furniture stores face costly returns and the kinds of deliveries that cannot be expedited.

(viii) **Corporate Vulnerability :** The availability of product details, catalogs and other information about a business through its web site makes it vulnerable to access by the competition. The idea of extracting business intelligence from the competitors Web pages is called **Web farming,** a term coined by Richard Hackathorn.

(ix) **Fulfillment Problems :** Tales of shipping delays, merchandise mix-ups and Web sites crashing under pressure continue to be a problem in e-tailing. Customer confidence in a e-commerce's ability to deliver during heavy shopping seasons continues to be a headache. Even happy customers says the experience could be improved.

(x) **System Scalability :** A business develops an interactive interface with customers via a Web site. After a while, statistical analysis determines whether visitors to the site one-time or recurring customers. If the company expects 2 million customers and 6 million show up, Web site performance is bound to experience degradation, slow down and eventually loss of customers. To keep this problem for happening, **a Web site must be scalable or upgradeable on a regular basis.**

(xi) **Consumer Search is not Efficient or Cost-efficient :** On the surface; the electronic marketplace appears to be a perfect market, where worldwide sellers and buyers share information and trade without intermediaries. However, a closer look indicates that new types of intermediaries are essential to e-commerce. They include electronic malls that guarantee product quality, mediators for bargaining and certificate authorities to ensure legitimacy of transactions. All these intermediaries add to transaction costs.

1.6. OPPORTUNITIES OF E-COMMERCE

We can divide the E-Commerce opportunities in three sections :

1. Commercial opportunities
2. Social Services opportunities
3. Entertainment opportunities.

1.6.1. Commercial Opportunities

E-commerce is suitable to apply in following types of commercial transactions :

(*a*) **Property Transactions :** Property Transactions can be executed by e-commerce facilities. Several estate dealers, builders, housing finance companies make use of e-commerce facilities and Internet sites to offer their products.

(*b*) **Sale, Purchase of Goods :** Consumers can buy products, industries can procure raw materials, components etc. manufactures and sellers can sell their products by using e-commerce.

(*c*) **Delivery of Goods :** Direct E-commerce permits the delivery of products. For example, the computer software is directly downloaded by the software manufacturer on computer of the customer.

(*d*) **Electronic Payments :** Payments can be made by utilizing e-commerce devices like credit card, debit card and electronic transfer etc.

(*e*) **Banking and Money Transfer :** Banking services are also carried out by e-commerce. Online banking, ATM facility and electronic money transfer are some of the means to carry out banking transaction through e-commerce.

(*f*) **Import and Export :** Electronic payments have played significant role in import-export business. The Internet has further simplified the import-export business. Importers can now make enquiries about the products, their manufacturers, price, other terms and conditions etc. Exporters can also make enquiries about suitable consumers. Payments can be made by electronic modes including digital means like Internet payment or internet money transfer.

(*g*) **Insurance Business :** Insurance business is the latest to utilize e-commerce facilities. Consumers can enquire about insurance products and can also subscribe to a policy through e-commerce media like fax, internet etc.

(*h*) **Transportation :** Transportation is crucial for handling and delivery of raw material, components, equipments and finished products. Facilities for carriage of goods by sea or air can be arranged and monitored by Internet facility.

(*i*) **International Finance :** Financial transactions involve both the person offering financial service as well those required it. Loans, Bills Discounting, Lease Transaction can be easily carried out by e-commerce devices. Avenues of International Finance can also be investigated by internet facilities.

(*j*) **Online share trading :** It is possible to sell, buy, pledge or transfer shares through e-commerce. Several companies, brokers offer this facility. Therefore, for dealing in securities, a customer is no more required to visit a broker.

(*k*) **Tourism and Hotel Industry :** It is possible to know about different tourist spots and accommodation throughout the world by surfing on the net. Customers can book their tickets, accommodation in hotels by using different e-commerce facilities like e-mail etc.

(*l*) **Online Depository Services :** National Stock Depository Limited and Central Stock Depository Limited and several other companies offer online depository services. Therefore, one can keep his securities in d-mat form by using e-commerce.

(*m*) **Payment of Taxes :** It is possible to make payment for taxes electronically.

(*n*) **Public Issue Subscription :** Public issue of securities can also be subscribed electronically by submitting the form and making payment through credit card.

(*o*) **Payment of Services :** Payment for different services like telephone, electricity, water, insurance premium, fee for children etc. can also be made by using e-commerce facilities.

(*p*) **Renewal of License :** Licenses can also be renewed by using e-mail or internet facilities.

1.6.2. Social Services

E-commerce can also be utilized for following social services :

(*a*) **Online Education and Distant Learning and Training :** E-commerce is equally suitable for imparting education. It is possible for the student to take lessons and imparting education. It is possible for the student to take lessons and interact with teachers through video conferencing facility. Students can also record lectures, download study material by use of different modes of e-commerce.

(*b*) **Online Diagnosis and Treatment :** Video conferencing has made it possible to consult a doctor through electronic devices. Even X-rays and surgical operations can also be performed through video conferencing.

(*c*) **Sharing research :** Research Scholars and Scientists sitting in different parts of the world can share each other's work and results through different modes of e-commerce.

(*d*) **Filing of tax return :** It is possible to file income tax return electronically. Finance Act, 2002-03 proposes such filling of return.

(*e*) **Statistics :** E-commerce helps to collect, analyse and interpret statistics relating to different matters.

1.6.3. Entertainment

E-commerce has its use in following areas of entertainment :

(*a*) **Ticketing for Theatres, Cinema Halls, Concerts :** Tickets can also be reserved for theatres and concerts electronically.

(*b*) **Online Games and Refreshers Courses :** E-commerce provides facilities for online games and refreshers courses.

1.7. TRENDS IN E-COMMERCE

There are numerous perspectives of electronic commerce which gives birth to new trends in business and society and these are discussed below :

1. From *historic perspective,* some authors see e-commerce as largely or entirely an internet phenomenon. Seddon has suggested that, "the world has just entered a third new phase in the evolution of IT capabilities; the internet era". The suggestion divides the evolution of information technology (IT) into 20 years periods:
   - **1955 – 1974** – The Electronic Data Processing (EDP) Era.
   - **1975 – 1994** – The Management Information System (MIS) Era.
   - **1995 – 2014** – The Internet Era.

2. From *an interface perspective,* e-commerce involves various information and transaction exchanges: business-to-business, business-to-consumer, consumer-to-consumer.
3. From a *business process perspective,* e-commerce includes activities that directly support commerce electronically by means of networked connections. Within business processes (manufacturing, inventorying, operation) and business-to-business process (supply chain management) are managed by the same networks as business-to-consumer processes.
4. From a *communications perspective,* e-commerce is the ability to deliver products, services, information or payments via networks such as the Internet and the World Wide Web.
5. From an *Online perspective,* e-commerce is an electronic environment that makes it possible to buy and sell products, services and information on the Internet. The products may be physical such as cars or services such as news and consulting.
6. From a *structural perspective,* e-commerce involves various media: data, text, web pages, Internet Telephony and Internet desktop video.
7. As a *market,* e-commerce is the world wide network. A local store can open a Web storefront and find the world at its doorstep-customers, suppliers, competitors and payment services.

SUMMARY

In this chapter we have established the basic platform to study e-commerce. Now after reading this chapter we are through with e-commerce basics. Now we can easily understand the e-commerce terminologies. In this chapter we also do the critical study of e-commerce and now we are aware of strength and weakness of e-commerce.

REVIEW QUESTIONS

1. "Now a days market is driven by e-commerce." Justify this statement.
2. What do you understand by E-commerce ? Discuss the driving forces for e-commerce.
3. If we want to setup the e-commerce business then what are the various required elements ?
4. What are the various strength and weakness of e-commerce.
5. Discuss the commercial, social as well as Entertainment Opportunities in e-commerce.
6. "Now a days e-commerce is over powered the traditional buying and selling system." Comment on this statement.
7. "Business becomes a global phenomenon through e-commerce." Justify this statement.

❑❑❑

2

CHAPTER

E-COMMERCE FRAMEWORK

INSIDE THIS CHAPTER

2.1. Electronic Commerce Framework

2.2. Types of Electronic Commerce

2.3. Intra/Inter Organizational E-Commerce

2.4. EDI (Electronic Data Interchange)

2.5. Electronic Funds Transfer (EFT)

2.6. Component of E-Commerce

2.1. ELECTRONIC COMMERCE FRAMEWORK

Basically to implement e-commerce, certain infrastructure is required and that infrastructure is a combination of computers, computer networks and communication softwares and finally information superhighway. Let us see the various activities of e-commerce and the infrastructure requirement to perform these activities of e-commerce :

(*i*) E-commerce provides buying and selling process so certainly e-commerce **infrastructure must support these common business services.**

(*ii*) We frequently sends and retrieve the information so there must be facility in **infrastructure for messaging and information distribution.**

(*iii*) The e-commerce related websites has the information about product and it is design so **infrastructure must have multimedia content and network publishing facilities.**

(*iv*) Finally **information superhighway is required** to performs the e-commerce transactions.

On the basis of above discussion we can say that e-commerce applications and infrastructure are two main pillers of e-commerce and they are indispensable :

(*i*) A clear cut government policy, public related issues, information access issues on internet, privacy of information, social and legal issues are the one piller of e-commerce.

(*ii*) Uniform technical standard for information publishing user interface, electronic documentation, multimedia and network protocol are the another piller of e-commerce.

Figure 2.1 make this concept more clear.

The successful e-commerce application will require the **I-way infrastructure in the same way as it required in the traditional business. The I-way is a mesh of interconnected data communication highways of many forms: telephone wires, cable TV wires, radio-based wireless-cellular and satellite.**

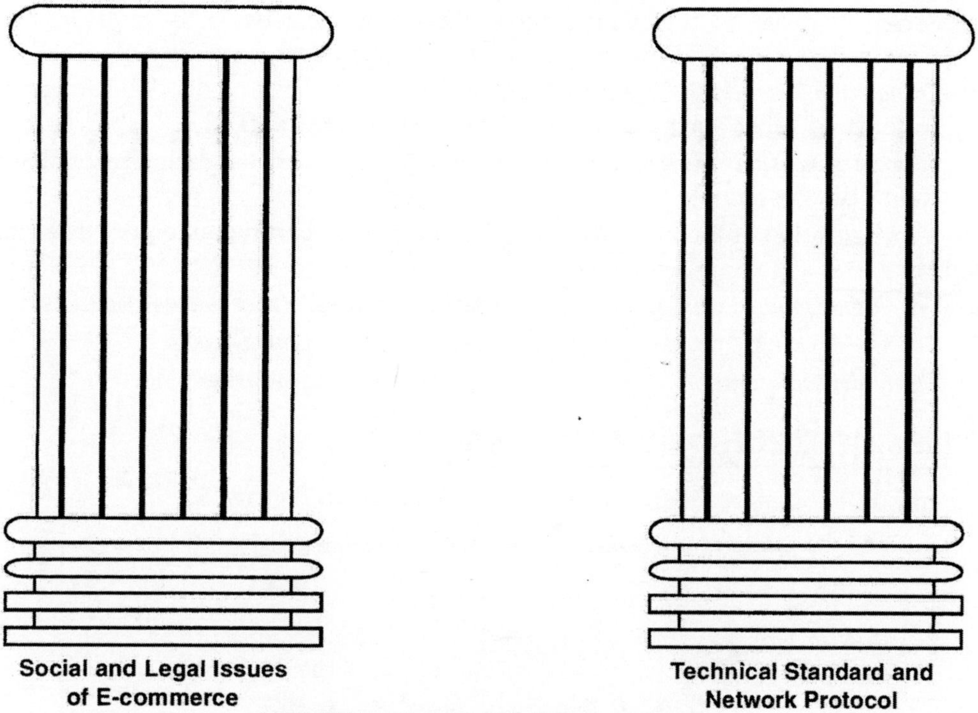

FIGURE 2.1. Framework for E-commerce.

In the electronic "highway system" the multimedia content is stored in the form of electronic documents, which are digitalized, compressed and stored in computerized libraries or multimedia storage warehouse referred as "SERVER" which are linked by transport networks to each other and to the hardware/software clients that allows the customers to access them.

On the I-way, messaging software fulfills the role of distribution in the form of e-mails, EDI or point-to-point file transfers. The encryption ad decryption authentication methods have been developed to ensure security of the contents while traveling the I-way and at their destination. The public policy issues deals with the cost of accessing the information, regulation to protect customers from fraud, intrusion to privacy rights, policing of global information traffic etc. The final pillar of e-commerce is technical standards to ensure harmonious integration across the network and access of information on any type of devices like laser discs, PCs portable hand held devices or television, set top boxes (cable converter box) and on all types of operating systems.

2.1.1. The Mechanism of Electronic Commerce

1. **The Business Dimension of E-commerce** : These are two bases and interactive business dimensions to Electronic Commerce and these are :
   (*a*) *The customer dimension* : This refer to the placing of finished goods and services with the final customers.
   (*b*) *The enterprise dimension* : This primarily an intercorporate or interorganizational (including business to government) supply chain management (SCM) etc.
2. **The Technological Dimensions of E-commerce :** It can be classified according to three basic function of any market environment.

(a) *Access environment* : It make use of private and public network technologies like Local Area Network (LAN), Wide Area Network (WAN), Internet etc.

(b) *Transaction dimension* : These are EDI, point of sale devices, credit, debit and smart card Automated Teller Machines and Electronic Fund Transfer (EFT).

(c) *Support dimension* : These are support services like card validation technologies, bar coding devices etc.

3. **The Configuration of E-commerce** : The EC to become operational require three things to happen :

(a) *The Organizational configuration* : Integrating Business Process Electronically

(b) *The network configuration* : Providing a backbone for e-commerce.

(c) *The media configuration* : Getting access to the electronic market place.

2.2. TYPES OF ELECTRONIC COMMERCE

We can divide e-commerce in four parts as shown below in Figure 2.2.

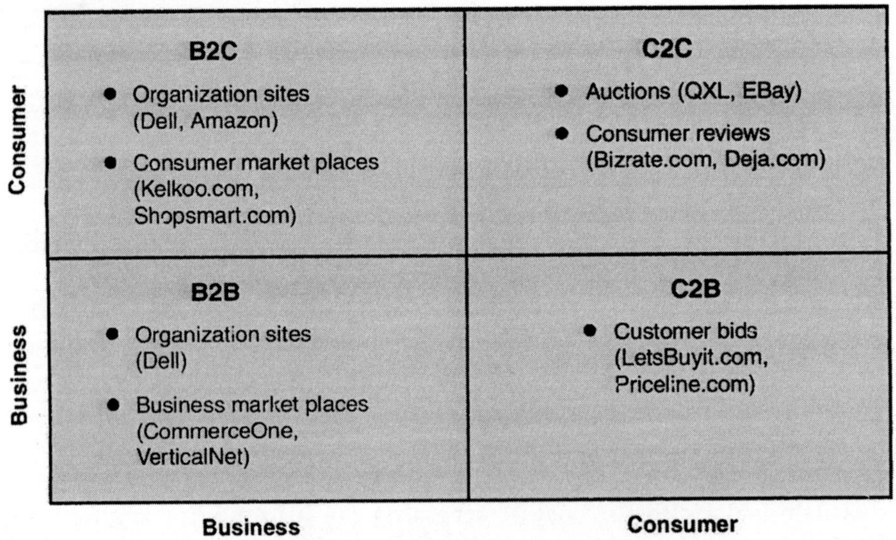

FIGURE 2.2. Summary of transaction alternatives between businesses and consumers.

2.2.1. Business-to-Customer Electronic Commerce

The B2C of electronic commerce transaction is ideally suited for the following types of merchandise.

(1) Goods that can be easily transformed into digital format, such as books, music, clips and videos and software packages.

(2) Items that follow standard specifications, like printer ribbons, ink, cartridges etc.

(3) Highly rated branded items or items with return security: such as Dell and Compaq computers, electronic gadgets from Sony etc.

(4) Items sold in packets that cannot be opened even in physical stores e.g. Kodak film rolls.

(5) Relatively cheap items where saving outweigh risks.

(6) Items that can be experienced online, such as music, videos etc.

2.2.2. Customer-to-Business Electronic Commerce

Customer-to-Business (C2B) can be described as a form of electronic commerce where the transaction originated by the customer has a set of requirement specifications and specific price for a commodity, service or item. It is the responsibility of electronic commerce business entity to match the requirements of the consumers to the best possible extent. Customer-to-Business (C2B) enables a consumer to determine the price of a product and/or service offered by a company.

2.2.3. Consumer-to-Consumer Electronic Commerce

Consumer-to-Consumer (C2C) is the electronic commerce activity that provides the opportunity for trading of products and/or services amongst consumers who are connected through the internet. In this category electronic tools and internet infrastructure are employed to support transaction between individuals. Traditional economic activities corresponding to 'classified advertisement' and auctions of personal possessions form the basis for the category. Much of the transaction in this category correspond to small gift items, craft merchandise and similar items that are normally sold through 'flea' markets or bazaars, where individuals sell their goods to other individuals at a market determined prices.

2.2.4. Business-to-Employee (B2E) Services

For an internal user, searching for a particular type of information from a vast information base of a company is a time consuming task. In knowledge based industries, human resource is the single most valuable asset, displaying itself in the performance of these organizations. Efficient management of the intellectual assets is crucial for creating better business value and gaining competitive advantage. The intranet based business to employee applications are often used for implementing improved employee relationship management initiatives. This business-to-employee (B2E) application offers employees a self-service capability many human resource functions.

2.3. INTRA/INTER ORGANIZATIONAL E-COMMERCE

An inter-organization involves information flow *among two or more Organizations.* Major objective is efficient and effective transaction processing like Transmitting orders, bills and payments using electronic data exchange.

"*Intra-organizational E-commerce* means intra business. It includes *all internal organizational activities* generally performed on intranets. It includes exchange of goods, services or information"

For example, an Institute or organization can set their own intranet and that can be used as Bulletin board or to send the organization related E-mails. That particular mail system is only for the intra organizational activities and not exposed to other persons.

2.3.1. Use of E-Mail in E-Commerce

Electronic mail, or Email, is the term used to describe the tool which allows one computer user to send a message to one or more other computer users over a computer network in a digital form. Email can be sent internally to members of an organization through their internal computer network or they can be sent externally to anyone in the world by using the network of computer networks that make up the internet.

It's very similar to our traditional mail system with post boxes, post offices, envelopes and addresses. The difference is that instead of India Post delivering our messages around the globe for us, messages are sent electronically to other computer users via the computer network (the internet) taking a fraction of the time to get there and, we don't have to walk to the letter box!

Just as we can send attachments with our letters in the normal post (snail mail) we can also send computer documents, graphics, software or anything else that can be turned into a digital form attached to an email message.

2.3.1.1. Intra-Organizational Mail

The mail deployed on the intranet of any organization is called intra-organizational mail. People working in that organization uses this mail system. They use this mail system to share the information between them.

2.3.1.2 Business to Customer E-Mail

This is the mail transaction that takes place between organization and customer. For example any marketing company which is doing some market survey, sends mails to the customers directly for collecting their feedback. This kind of e-mail transaction comes under B2C mail.

2.3.1.3 Business to Business E-Mail

This is the mail transaction that takes place between two organizations. For example any organization sends mail to another organization for some business query. This kind of e-mail transaction comes under under B2B mail.

2.3.1.4 Customer to Customer E-Mail

This is the mail transaction that takes place between different customers. For more then one customer shares mail about any organization to do some query about the product of that company. This kind of e-mail transaction comes under C2C mail.

2.4. EDI (ELECTRONIC DATA INTERCHANGE)

Electronic Data Interchange (or EDI) refers to the use of computers in formatting and exchanging business information using internationally recognized standards and formats. Standards organizations throughout the world have defined hundreds of document types to create a universal language for companies to conduct business while using dissimilar systems and processes. Common business documents, such as the invoice, purchase order, production planning schedule, ship manifest, and customs declaration, have been standardized and are sent and received electronically over private secure networks (Value-Added Networks—VANs) and the Internet.

The **prominent use of EDI** continues to change many traditionally paper-based business procedures. Increasingly, improvements in technology have made EDI available for small and mid-sized companies to easily participate and take advantage of business benefits typically seized by larger members of the supply and demand chain. Large companies continue to expand the scope of information exchanged with suppliers and customers and rely on EDI to increase visibility, efficiency and response to market conditions.

2.4.1. EDI Over Wan/Internet/Extranet

EDI can also be used in exchanging financial information and payments in electronic form. The electronic fund transfer (EFT) systems used by financial institutions are a prime example of the application of EDI in the banking and financial sector. EDI should be viewed as a mean to streamline procedure and improve efficiency and productivity.

According to the Data Interchange Standards Association, "Electronic Data Interchange (EDI) is the Computer-to-Computer exchange of business data in standard formats. In EDI, information is organized according to a specified format set by both parties, allowing a "handsoff" computer transaction that requires no human intervention or rekeying on either end. All information contained in an EDI transaction set is, for the most part the same as on conventionally printed document."

The figure given below depicts EDI over internet, extranet, and wide area network and Electronic Banking in a country is good example of this concept. Person can do the financial transaction online for a particular bank, which have network over wan.

It is very clear from Fig. 2.3 that when we talk about :

- **Intranet :** Transacting at organization level only.
- **Extranet :** Transacting between company, supplier, and customers.
- **Internet :** Transaction can take place with any where with any body in the world.

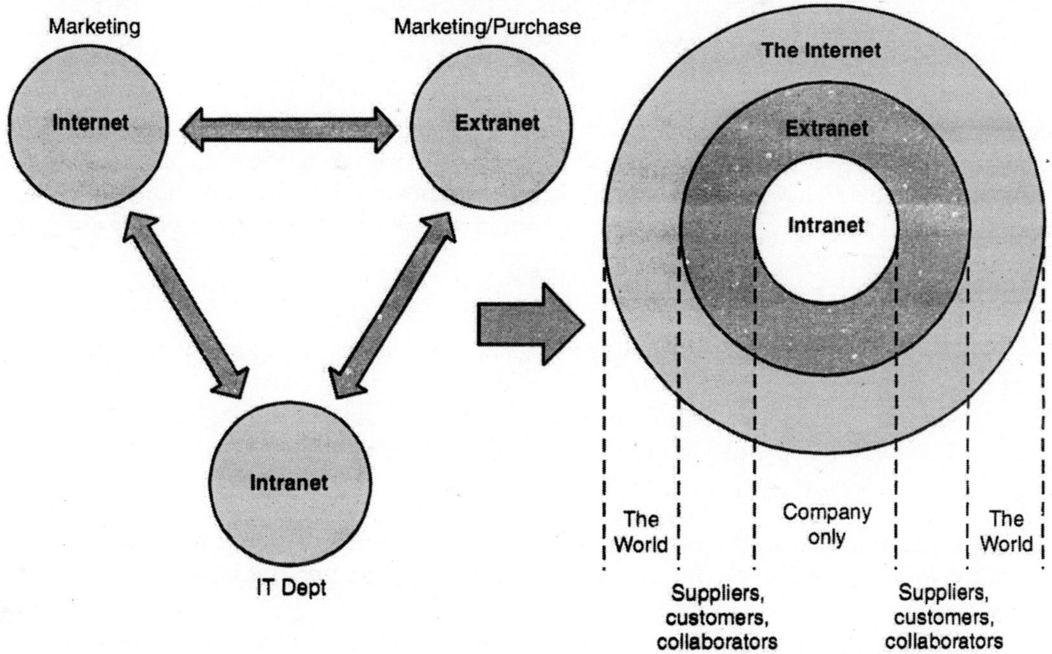

FIGURE 2.3. The relationship between intranets, extranets and the internet.

2.5. ELECTRONIC FUNDS TRANSFER (EFT)

Electronic Funds Transfer (EFT) is a system of transferring money from one bank account directly to another without any paper money changing hands. One of the most widely-used EFT programs is Direct Deposit, in which payroll is deposited straight into an employee's bank account, although EFT refers to any transfer of funds initiated through an electronic terminal, including credit card, ATM and point-of-sale (POS) transactions. It is used for both credit transfers, such as payroll payments, and for debit transfers, such as mortgage payments.

The bank through the Automated Clearing House (ACH) network, the secure transfer system that connects all India's financial institutions, processes transactions. For payments, funds are transferred electronically from one bank account to the billing company's bank, usually less than a day after the scheduled payment date.

The growing popularity of EFT for online bill payment is paving the way for a paperless universe where checks, stamps, envelopes, and paper bills are obsolete. The benefits of EFT include reduced administrative costs, increased efficiency, simplified book-keeping, and greater security. However, the number of companies who send and receive bills through the Internet is still relatively small.

2.6. COMPONENT OF E-COMMERCE

We can divide E-Commerce in three component :

(*i*) Institutions (*ii*) Process (*iii*) Networks

2.6.1 Institutions of E-Commerce

Institutions plays major role in the success of e-commerce. Institutions like government, merchants, manufacturers, suppliers, consumers, banks, and financial institutions are the key to implement to e-commerce. Let us discuss their role in short :

1. Government

Despite the importance of technological and skill infrastructures, it is the politics of Governance initiatives that probably hold the key. E-commerce projects have made slow progress in many countries because they do not serve the political self-interests of the major stakeholders, particular senior public officials and government. Government is responsible to make the policy and rules of e-commerce in any country.

2. Merchants

As we know that every institute involve in e-commerce must be equally competitive for the success of the e-commerce. Merchants are very essential part of any type of business, they must show willingness and courage to implement e-commerce in their traditional business. Merchants holds the key for the success of the e-commerce since they are going to invest the money.

3. Suppliers

Suppliers are equally important component of e-commerce, they must have suitable infrastructure to support the e-commerce then only they can get the benefit of e-commerce technology.

4. Consumers

We know that consumer are key of any business so consumers must we aware about the latest development in e-commerce. It is very essential to change the mind set of consumers. So that they can start using the e-commerce then only it will be successful.

5. Banks

Nowadays banks are rapidly going for e-commerce to make their working efficient. Automated Teller Machines or 24-hour Tellers are electronic terminals that let our bank almost any time. To withdraw cash, make deposits, or transfer funds between accounts, we generally insert an ATM card and enter your PIN. Some financial institutions and ATM owners charge a fee, particularly to consumers who don't have accounts with them or on transactions at remote locations. Generally, ATMs must tell us they charge a fee and its amount on or at the terminal screen before we complete the transaction. Check the rules of your institution and ATMs you use to find out when or whether a fee is charged.

6. Financial Institutions

Other non-banking institutions are also adopting e-commerce, for example now we can submit LIC premium on line from anywhere.

2.6.2. Process of E-Commerce

The successful implementation of the e-commerce depends on perfect processes of the e-commerce. Some very important part of the process are as follows :

(*i*) Marketing (*ii*) Sales

(*iii*) Payment (*iv*) Support and Service.

The complete business is the game of the marketing now a days, so marketing process **should be compatible to e-commerce. Now more and more marketing becomes on-line and web base.**

Sales are **another important department of business activities so it must support e-commerce.**

Now a days mode of the payment becomes electronic. **Mostly transaction take place on line through Cyber cash. Merchants are tie-up with banks for online payment, credit and debit card payments. After sale** service and support **is the backbone of any business so more and more companies using the e-commerce to provide this service through telephone, mail and internet.**

SUMMARY

In this chapter we learn the basic requirement of e-commerce. A specific infrastructure is required to implement the e-commerce. We also study various kind of e-commerce like B2C, B2B and C2C. The efficiency of organization can be increased by using e-commerce tools. E-commerce can be implemented through intranet, extranet and Internet. Institutions, Process and networks are basic component of e-commerce.

REVIEW QUESTIONS

1. What is basic framework requirement of e-commerce implementation ?
2. Write the short notes on the following :
   (*i*) B2B e-commerce.
   (*ii*) C2C e-commerce.
   (*iii*) B2C e-commerce.
   (*iv*) Intra organizational e-commerce.
3. What is difference between e-commerce over extranet and intranet ?
4. What are the basic component of e-commerce ?
5. How EFT is helpful in banks ?
6. How EDI can change the life of common man ?
7. How the successful implementation of e-commerce depends on the processes of e-commerce.

❑❑❑

3

CHAPTER

ARCHITECTURAL FRAMEWORK OF E-COMMERCE

INSIDE THIS CHAPTER

3.1. Internet
3.2. World Wide Web
3.3. TCP/IP Protocols
3.4. Application Level Protocols
3.5. Web Browser
3.6. Web Servers
3.7. Common Gateway Interface
3.8. Hypertext Markup Language (HTML)
3.9. Electronic Data Interchange
3.10. Transaction

3.1. INTERNET

The Internet refers to millions of computers connected to a gigantic network and communicating via **TCP/IP protocols**.

A protocol is a predefined way for a computer to communicate with another computer, for instance when requesting a service, or when forwarding some information to another machine. Each computer at any given time has a unique address on the Internet. **This is its IP address.**

3.1.1. Internet History

Until the 1960s, very little communication was taking place between computers, and was carried out by the technology of telephone networks, i.e. circuit switching. In 1962 Paul Baran and Donald Davies independently proposed an idea of a robust, efficient, store-and-forward data network based on packets, *i.e.* units of data carried independently from one another. The technology was called **packet switching**, and has been implemented as ARPANET (a network of research sites in US, a project funded by the US Department of Defense) in late 1960s. In early 1970s, ARPANET spanned the continental US, and by 1973 it had connections to Europe.

ARPANET had several protocols for communications of computers one to another or with the network. In 1974, a new, more robust suite of protocols has been developed and implemented on the ARPANET based on **Transmission Control Protocol** (TCP) and **Internet Protocol** (IP). The standard includes a large collection of protocols, some of which we mention later. These protocols have been modified several times since then, but essential ideas of the original protocols are still preserved. In 1983, the US Department of Defense has mandated that all of their computer systems use TCP/IP protocols, which boosted use of these protocols in the US and throughout the world.

Another boost to TCP/IP was given by including it, also in 1983, into the communication kernel of the University of California's UNIX implementation.

In 1986, the National Science Foundation (NSF) built a backbone network (i.e. a network of fast powerful computers which can quickly forward information to each other and other computers) called NSFNET. Eventually it grew to provide connectivity between various networks, forming the foundation of what currently is known as Internet.

In 1993 the NSF started to reduce its role in governing the Internet. Currently the administration of the Internet is divided between several international organizations, such as ISOC (Internet Society) and IAB (Internet Activities Board), which supervise various technical, administrative, and other aspects of the Internet.

Modern Internet has over 32,000,000 registered domain names (according to domainstats.com/ provided by ISOC). The size of Internet doubles every 10-12 months.

3.1.2. Uniform Resource Locator (URL)

URL is an acronym for *Uniform Resource Locator* and is a reference (an address) to a resource on the Internet.

The following is an example of a URL which addresses the Java Web site hosted by Sun Microsystems:

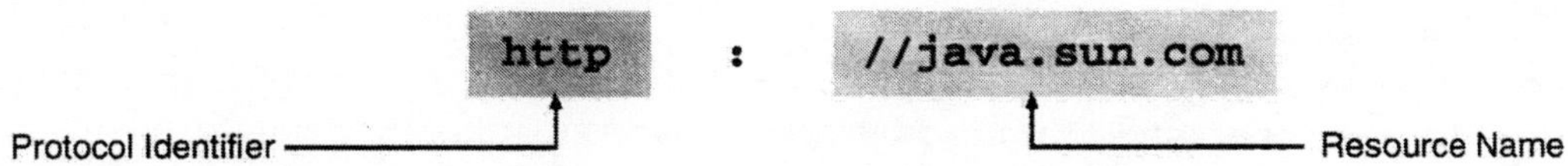

As in the previous diagram, a URL has two main components :

- Protocol identifier
- Resource name

Note that the protocol identifier and the resource name are separated by a colon and two forward slashes. The protocol identifier indicates the name of the protocol to be used to fetch the resource. The example uses the Hypertext Transfer Protocol (HTTP), which is typically used to serve up hypertext documents. HTTP is just one of many different protocols used to access different types of resources on the net. Other protocols include File Transfer Protocol (FTP), Gopher, File, and News.

The resource name is the complete address to the resource. The format of the resource name depends entirely on the protocol used, but for many protocols, including HTTP, the resource name contains one or more of the components listed in the following table :

Host Name	The name of the machine on which the resource lives.
Filename	The pathname to the file on the machine.
Port Number	The port number to which to connect (typically optional).
Reference	A reference to a named anchor within a resource that usually identifies a specific location within a file (typically optional).

For many protocols, the host name and the filename are required, while the port number and reference are optional. For example, the resource name for an HTTP URL must specify a server on the network (Host Name) and the path to the document on that machine (Filename); it also can specify a port number and a reference. In the URL for the Java Web site java.sun.com is the host name and the trailing slash is shorthand for the file named/index.html.

3.1.3. IP Addresses

Every machine on the Internet has a unique identifying number, called an IP Address. The IP stands for Internet Protocol, which is the language that computers use to communicate over the Internet. A protocol is the predefined way that someone who wants to use a service talks with that service. The "someone" could be a person, but more often it is a computer program like a Web browser.

A typical IP address looks like this:

216.27.61.137

To make it easier for us humans to remember, IP addresses are normally expressed in decimal format as a dotted decimal number like the one above. But computers communicate in binary form. Look at the same IP address in binary.

11011000.00011011.00111101.10001001

The four numbers in an IP address are called octets, because they each have eight positions when viewed in binary form. If we add all the positions together, we get 32, which is why IP addresses are considered 32-bit numbers. Since each of the eight positions can have two different states (1 or zero), the total number of possible combinations per octet is 2^8 or 256. So each octet can contain any value between zero and 255. Combine the four octets and you get 2^{32} or a possible 4,294,967,296 unique values.

Out of the almost 4.3 billion possible combinations, certain values are restricted from use as typical IP addresses. For example, the IP address 0.0.0.0 is reserved for the default network and the address 255.255.255.255 is used for broadcasts.

The octets serve a purpose other than simply separating the numbers. They are used to create classes of IP addresses that can be assigned to a particular business, government or other entity based on size and need. The octets are split into two sections: Net and Host. The Net section always contains the first octet. It is used to identify the network that a computer belongs to. Host (sometimes referred to as Node) identifies the actual computer on the network. The Host section always contains the last octet. There are five IP classes plus certain special addresses

3.1.4. Domain Names

A name that identifies one or more *IP addresses* is called domain name. For example, the domain name *microsoft.com* represents about a dozen IP addresses. Domain names are used in URLs to identify particular Web pages. For example, in the URL *http://www.kiet.edu/index.html,* the domain name is kiet.edu.

Every domain name has a suffix that indicates which top level domain it belongs to. There are only a limited number of such domains. For example:

- gov - Government agencies
- edu - Educational institutions
- org - Organizations (nonprofit)
- mil - Military
- com - Commercial business
- net - Network organizations
- ca - Canada
- th - Thailand
- in - India

Because the Internet is based on IP addresses, not domain names, every Web server requires a Domain Name System (DNS) server to translate domain names into IP addresses.

3.1.4.1. Domain Name System

DNS stands for two things: Domain Name Service and Domain Name Servers. One acronym defines the protocol, the other defines the machines that provide the service. The job that DNS performs is very simple: it takes the IP addresses that computers connected to the Internet use to communicate with each other and it maps them to hostnames.

Human beings tend to have a difficult time remembering long strings of seemingly arbitrary numbers. The way that our brains work, it's difficult to make information like that stick. And that is where DNS comes in. It allows us to substitute words or phrases for those strings of numbers. Words are a lot easier for people to remember than numbers, especially when they can be tied to a specific idea that is linked to the website.

DNS translates IP addresses into hostnames and back again. The hostnames are for the benefit of human end users. The IP addresses are the only essential thing, as far as the computers are concerned. In a longer form, we need to begin by looking at the different types of DNS servers.

3.1.5. Internet Infrastructure and Routing

Suppose we want to send a message (some piece of data) to another computer. Given the computer's IP address, how does the data find its way on the Internet? The information used to get data to its destination is contained in routing tables kept by each router, i.e. a machine that routes packages. Each router knows about it's sub-networks and which IP addresses they use. The router usually doesn't know what IP addresses are in other subnetworks of the Internet. Routers form a tree-like structure on the Internet with NSP (Network Service Provider) backbones at the roots, connected to one another.

When a packet arrives at a router, the router examines the IP address of its destination. The router checks it's routing table. If the network containing the IP address is found, the packet is sent to that network. If the network containing the IP address is not found, then the router sends the packet on a default route, usually **up** the backbone hierarchy to the next router. Hopefully the next router will know where to send the packet. If it does not, again the packet is routed upwards until it reaches a NSP backbone. The routers connected to the NSP backbones hold the largest routing tables and here the packet will be routed to the correct backbone, where it will begin its journey 'downward' through smaller and smaller networks until it finds its destination.

This process is known as **package routing**, where a package is a piece of data wrapped into an "envelope" with all the necessary information for routing the package and for sending it to the corrected application on the destination machine. TCP/IP protocols described below provide the wrapping and unwrapping.

3.1.6. Client/Server Model of Computing

To truly understand how much of the Internet operates, including the Web, it is important to understand the concept of client/server computing. The client/server model is a form of distributed computing where one program (the client) communicates with another program (the server) for the purpose of exchanging information.

The client's responsibility is usually to :

1. Handle the user interface.
2. Translate the user's request into the desired protocol.
3. Send the request to the server.

4. Wait for the server's response.
5. Translate the response into "human-readable" results.
6. Present the results to the user.

The server's functions include :

1. Listen for a client's query.
2. Process that query.
3. Return the results back to the client.

A typical client/server interaction goes like this :

1. The user runs client software to create a query.
2. The client connects to the server.
3. The client sends the query to the server.
4. The server analyzes the query.
5. The server computes the results of the query.
6. The server sends the results to the client.
7. The client presents the results to the user.

Repeat as necessary.

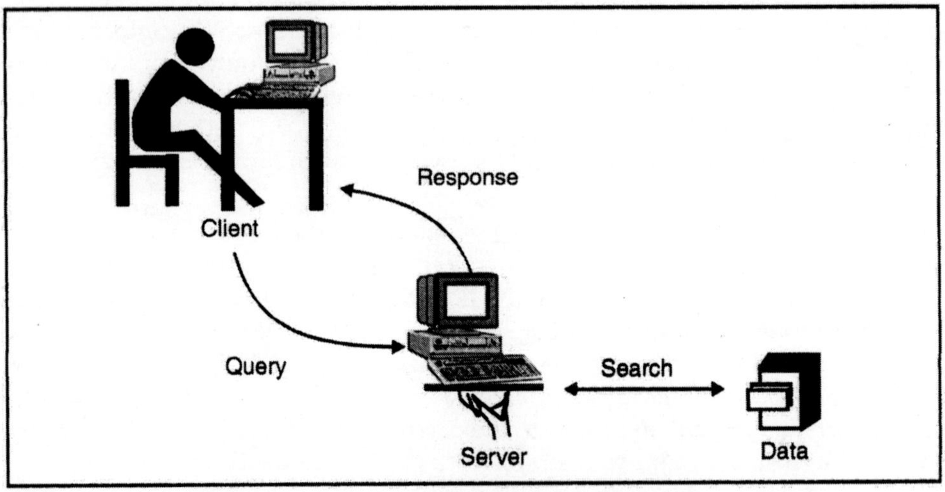

FIGURE 3.1. A typical client/server interactions.

This client/server interaction is a lot like going to a French restaurant. At the restaurant, we (the user) are presented with a menu of choices by the waiter (the client). After making our selections, the waiter takes note of our choices, translates them into French, and presents them to the French chef (the server) in the kitchen. After the chef prepares our meal, the waiter returns with our diner (the results). Hopefully, the waiter returns with the items we selected, but not always; sometimes things get "lost in the translation."

Flexible user interface development is the most obvious advantage of client/server computing. It is possible to create an interface that is independent of the server hosting the data. Therefore, the user interface of a client/server application can be written on a Macintosh and the server can be written on a mainframe. Clients could be also written for DOS- or UNIX-based computers. This allows information to be stored in a central server and disseminated to different types of remote computers. Since the user interface is the responsibility of the client, the server has more computing resources to spend on analyzing queries and disseminating information. This is another major advantage of client/server computing; it tends to use the strengths of divergent computing platforms to create more powerful applications. Although its computing and storage capabilities are dwarfed by those of the mainframe, there is no reason why a Macintosh could not be used as a server for less demanding applications. In short, client/server computing provides a mechanism for disparate computers to cooperate on a single computing task.

3.2. WORLD WIDE WEB

World wide web is broad term, which is the collection of distributed documents referred to as "pages" located on computers (or servers) all over the world. For example : http://kiet.edu.

Servers store hypertext markup language (HTML) files and respond to requests. Through the use of a browser, PC users can find and view server-based documents. Browsers like Netscape ensure easy access to server-based documents and display multimedia data. Future browser will include editor-like abilities, application linking, audio and video integration; data based front end, and ways to combine information with transactions.

Reason for the amazing success of the Web include :

1. Ease of navigation and use.
2. Ease of publishing content.
3. New distribution models.
4. Enabling a network-centric computing paradigm.
5. Enabling new intra-business applications.

3.2.1. Web Architecture

To use the Web, in addition to an Internet connection, a user need a special piece of software called a Web browser (such as Netscape Navigator). The browser act as a graphical interface between the user and the Internet – it sends the necessary command to request data from other computers and then format them for the user's screen. Documents that are formatted using hypertext markup language (HTML) contain taglines that inform the browser how to format them.

The web is based on a three-part architecture :

1. **HTML :** Hypertext markup language, the format for web pages, provide both formatting and hyperlinking.
2. **HTTP :** Hypertext transfer protocol, the protocol for communications between Web servers and browsers.
3. **CGI :** The common gateway interface, the interface for invoking programs from web servers.

Figure 3.2 shows the component of web architecture, including the web clients, web servers, HTTP protocol language for Web client/server communication and applications. Within the confines of this architecture, hundreds of entrepreneurs have produced tidal wave of software programs and tools.

The web architecture forms the basis for electronic commerce applications that involve building software in which the functions are distributed among :

1. Application servers (where application reside),
2. Data servers (where most of the data resides), and
3. A Group of client computers that are usually networked PCs (where the information user work).

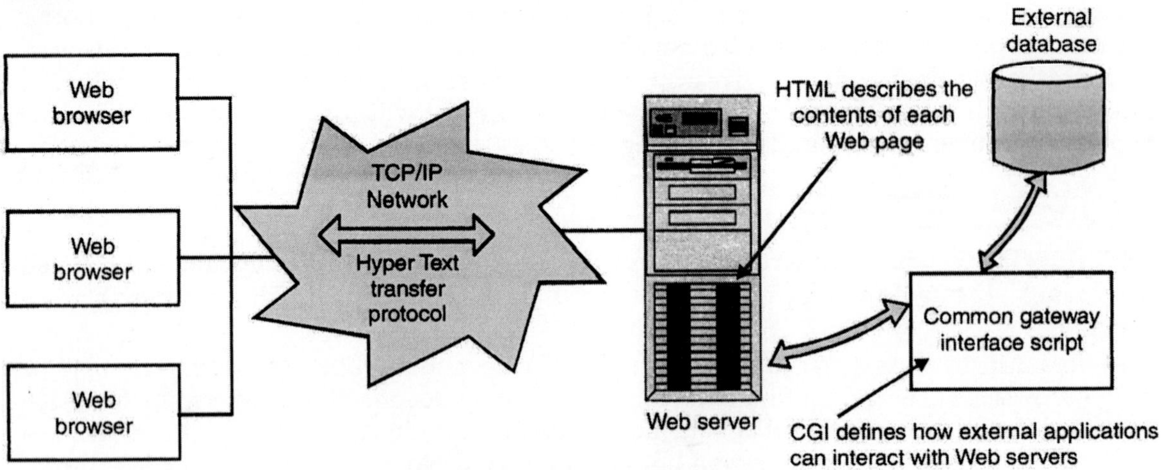

FIGURE 3.2. The web architecture.

3.2.2. The Web and E-Commerce

Figure 3.3 shows a block diagram depicting the key elements that constitute a web-based electronic commerce architecture: client browser, web server and third-party services. The client browser interacts with the web server, which then intermediates the action with third party services. The web server functions can be categorized into information retrieval, data and transaction management and security. The third party services could be other web servers that provide content, information processing tools and electronic payment systems.

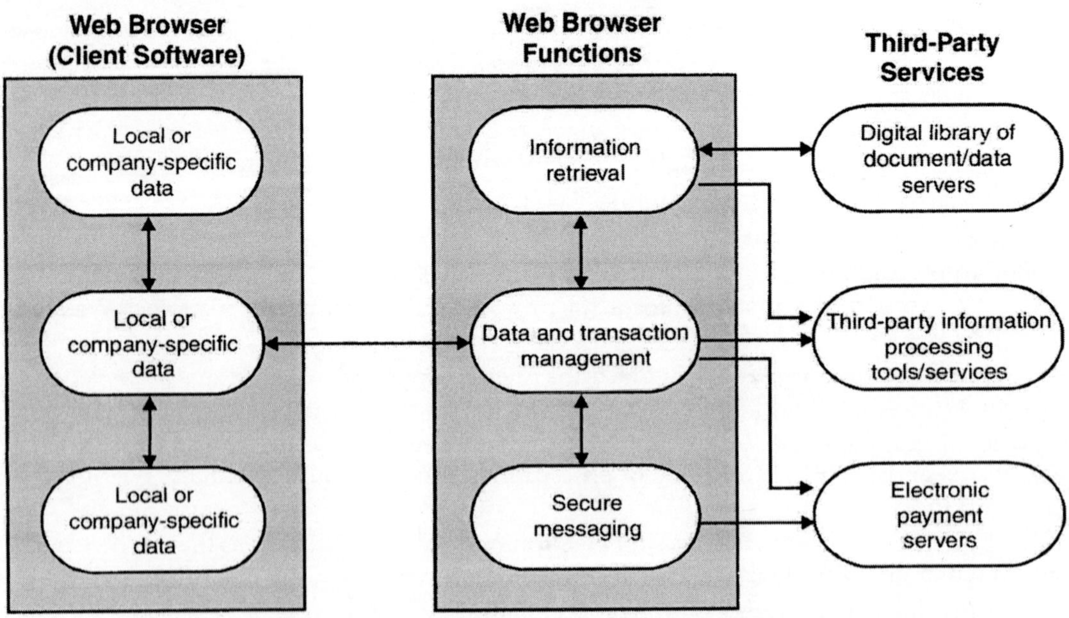

FIGURE 3.3. Block diagram depicting an electronic commerce architecture.

The importance of web for e-commerce can be understood by following :

(*i*) Companies use the Web to communicate with customers and suppliers by publishing content on their web server for widespread distribution. Motivated by the potential for business-to-business as well as business-to-consumer commerce, many firms are taking steps towards selling their products on the Web.

(*ii*) The Web is also changing and reshaping industries whose core business is information transfer between the firm and the consumer. Examples of such business functions include advertising, marketing, sales and customer service. Industries that are engaged in information transfer with customer include banking and financial services, retailing and electronic publishing and entertainment.

Within these industries, the Web is used for four major tasks :

1. Attracting new customers via marketing and advertising.
2. Serving existing customers via customer service and support function.
3. Developing new markets and distribution channels for existing products.
4. Developing new information based products.

3.3. TCP/IP PROTOCOLS

We know that rules are very important to make the communication possible and TCP/IP are the protocols which make the communication possible over internet.

There are 4 protocol layers that data goes through before it gets sent off to another computer. If the message to be sent is long, it will be broken into smaller chunks called **packets**.

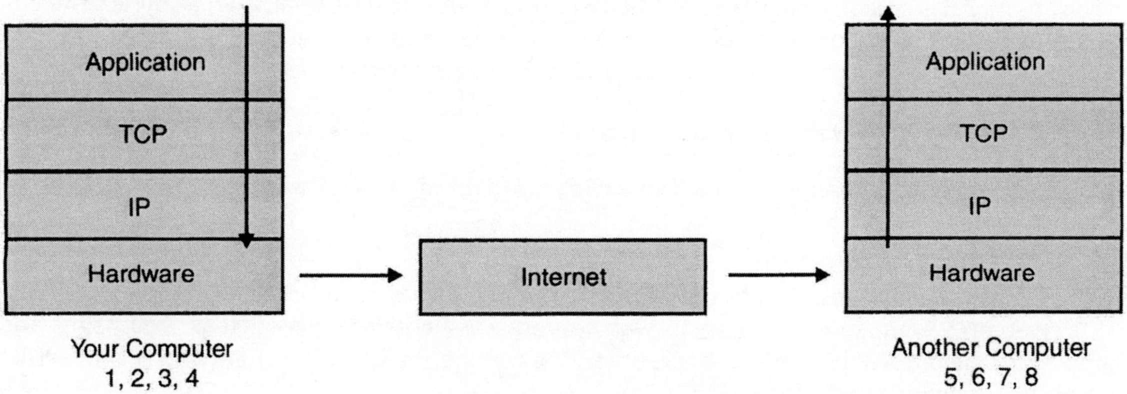

FIGURE 3.4. TCP/IP Protocol.

1. **Application Protocols Layer** : Protocols specific to applications such as WWW, e-mail, FTP, etc.
2. **Transmission Control Protocol Layer** : TCP attaches a **port number** of the destination service to packets to be sent out, and directs arriving packets to a specific application on a computer using a port number. Every internet service communicates at a specific port. Ports are integer numbers which allow a computer to distinguish between several internet communications happening at the same time, but facilitates by different internet applications. For instance, since FTP and HTTP use different ports, a computer can be downloading a file and displaying web pages at the same time. Most common ports are :
   - FTP : 20/21
   - Telnet : 23
   - HTTP : 80

TCP also handles assembly and disassembly of long messages into packets (smaller chunks of data). Each packet will be "wrapped" separately by all the protocols that the packet goes through. It will also get a sequence number, so that the packets will be reassembled at the destination (packets may arrive out of order). The last packet is marked to indicate that the message is complete.

3. **Internet Protocol Layer :** IP converts the domain name into IP address, attaches the IP address of the sending machine, directs packets to a specific computer using an IP address.
4. **Hardware Layer :** Converts binary packet data to network signals and back (For example, ethernet network card, modem for phone lines, etc.). That's how the data actually gets transmitted.

TCP is a Connection-Oriented, Reliable, Byte Stream Service : Connection-oriented means that two applications using TCP must first establish a connection before exchanging data. TCP is reliable because for each packet received, an acknowledgement is sent to the sender to confirm the delivery. TCP also includes a checksum in it's header for error-checking the received data. It sends data in byte streams, not character streams, meaning that any binary data (pictures, Excel files, etc.) may be sent using TCP.

Unlike TCP, **IP is an unreliable, connectionless protocol.** IP doesn't care whether a packet gets to it's destination or not. Nor does IP know about connections and port numbers. IP's job is too send and route packets to other computers. IP packets are independent entities and may arrive out of order or not at all. It is TCP's job to make sure packets arrive and are in the correct order.

The Figure 3.3 is actually greatly simplified: both TCP and IP are actually not a single protocol, but a suite of small protocols, each performing small tasks, such as establishing connection with another machine, checking packages for corrupted data, and so on.

3.4. APPLICATION LEVEL PROTOCOLS

Application level protocols, s.a. HTTP or FTP, build on top of TCP and IP.

3.4.1. HTTP

HTTP is a connectionless text based protocol to make the communication possible over web. Clients (web browsers) send requests to web servers for web elements such as web pages and images. After the request is serviced by a server, the connection between client and server across the Internet is disconnected.

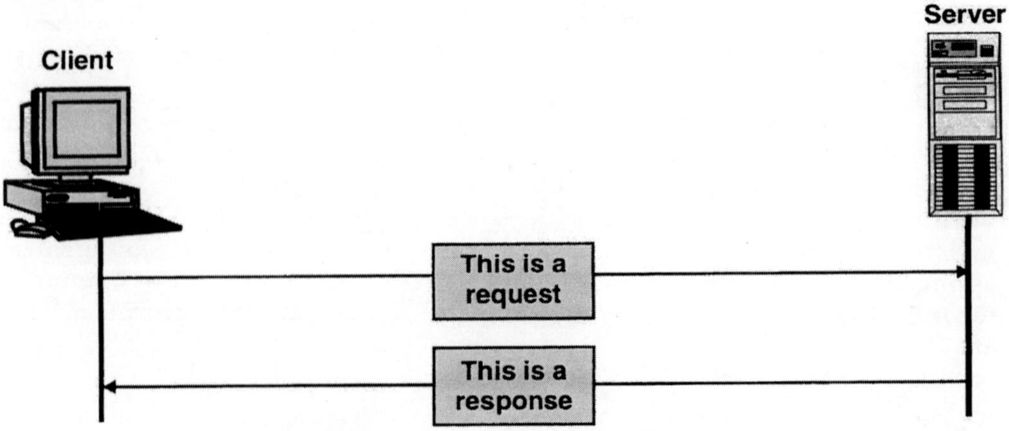

FIGURE 3.5. HTTP request and response.

A new connection must be made for each request. This means that HTTP is not connection-oriented. On the practical side, it means that after a requested HTML file gets downloaded, your computer is disconnected from the server, then your browser discovers that it needs a .gif image for the page, makes another request to the server, disconnects again and so on.

Each HTTP interaction works in following ways :

1. The browser established a connection with the server.
2. The browser send a request to server.
3. The server sends response to the client (browser).
4. Both browser and server disconnect and transaction ends.

3.4.2. Telnet Protocol

Telnet is method of connecting to a remote computer over the Internet that let us use programs and data just as if we were using the computer locally. Telnet is a text-only protocol. At one time it was one of the most common ways to connect to other sites. Now with the growth of the web, it is hardly used. There are still a few sites and resources that can only be reached through telnet.

Use of Telnet

- If we are a gamer, there are a number of text based games still available through telnet. See Yahoo's page on MUDs (Multi-User Dungeons).
- Outside of games, unless we already know of a particular use, such as connecting to a particular government database, we probably have no use for it.
- For years, many library catalogs were only reachable through telnet. Many early community resources, called free-nets, were reachable through telnet. You would have trouble finding any of these now.

Connect to a Telnet Address

If our Internet provider supports telnet (not all do), we can connect to a telnet address in one of two ways :

- Recent versions of Windows include a command line telnet. Begin at Start/Run and type **command**. On the next screen, type your telnet command. From a SLIP or PPP or ethernet connected account: type a telnet URL into the location window of our browser. To do this, our browser must be configured with a supporting version of telnet software. If we are using Windows95, make sure we get a version that supports Win95's 32-bit stack.
- From a unix shell account or other account that lets you type telnet commands, we can issue a command like this: **telnet some.address port-number**

3.4.3. File Transfer Protocol

FTP is means of communication over the network specifically designed for moving files between computers on the network. Primary use to most of us is transferring files to/from our desktop/laptop personal computers and the unix workstations, but it can also be used to communicate with machines at companies like Netscape to download software, or any PCs or other machines running FTP servers.

FTP servers listen on port 21. Data connections are initiated by the server from its port 20, to a port on the client identified in a PORT command. Following are recommended Programs for FTP:

Mac:	Fetch	Nice graphical interface that supports drag-and-drop
PC:	WS-FTP	Pretty good graphical interface once you get used to it.
	built-in	Windows 95 built-in support for ftp, command-line oriented.
Unix:	ftp	Standard Unix command-line ftp program, virtually identical to the PC built-in.
	ncftp	Slightly more feature-packed version, also command line oriented.

Using FTP to Download (Transfer from Another Machine to Yours)

1. **Connect to the Server (and Login) :** Servers with restricted access require a username and a password. Other servers (such as ftp.netscape.com) support "anonymous login" where the standard is to use the username "anonymous" and our full email address as your password.
2. **Find the File(s)/Folder(s) We Want to Download :** Navigating the folder hierarchy is just like it is on Macs, PCs, etc. Double-click a folder to open it. (Command-line FTP programs are more like navigating in dos/unix; use the "cd" command).
3. **Choose the Appropriate Download Mode :** There are two transfer modes in FTP. The first is called either "ascii" or "text", and the other is called "binary" or "image". The reason for the two modes is that different machines represent End-Of-Line characters in different ways. "text" mode automatically corrects for these differences so that the text we download appears correct on our machine, so for text files we should use text mode (otherwise, we may find, for example, that all our text appears on one line after it is downloaded). But if we use text mode on a non-text file, it will corrupt the file. It will try to interpret non-text data as text and therefore change the file. And for something non-text like a graphic or a microsoft word document (which contains formatting), you will find very unpleasant results if you use text mode! In short, if it isn't text, use binary mode (which leaves the file untouched).
4. Begin the transfer and get the file (make sure we know where the file is being put on our local machine).

Using FTP to Upload (Transfer from Your Machine to Another)

1. **Connect to the Server (and Login) :** Servers with restricted access require a username and a password. Other servers (such as ftp.netscape.com) support "anonymous login" where the standard is to use the username "anonymous" and our full email address as our password.
2. If we are using Fetch, choose to "Put" files/folders, and we will be prompted to specify the files.
3. Find the file(s)/folder(s) we want to upload on your computer. Navigating the folder hierarchy is just like it is on Macs, PCs, etc. double-click a folder to open it.
4. **Choose the Appropriate Upload Mode!** See the notes on download for details. Note that Fetch chooses different names for the file types when uploading. "Raw Data" is the same as binary/image. "Text" is text.
5. Begin the transfer.

3.4.4. The SMTP Protocol

SMTP (Simple Mail Transfer Protocol) is a TCP/IP protocol used in sending and receiving e-mail. However, since it's limited in its ability to queue messages at the receiving end, it's usually used with one of two other protocols, POP3 or Internet Message Access Protocol, that let the user save messages in a server mailbox and download them periodically from the server. In other words, users typically use a program that uses SMTP for sending e-mail and either POP3 or IMAP for receiving messages that have been received for them at their local server.

SMTP usually is implemented to operate over Transmission Control Protocol port 25. Simple Mail Transfer Protocol (SMTP), is Internet's standard host-to-host mail transport protocol and traditionally operates over TCP, port 25. In other words, a UNIX user can type telnet hostname 25 and connect with an SMTP server, if one is present.

If mail delivery fails, sendmail (the most important SMTP implementation) will queue mail messages and retry delivery later. However, a backoff algorithm is used, and no mechanism exists to poll all Internet hosts for mail, nor does SMTP provide any mailbox facility, or any special features beyond mail transport. For these reasons, SMTP isn't a good choice for hosts situated behind highly unpredictable lines (like modems). A better-connected host can be designated as a DNS mail exchanger, then arrange for a relay scheme. Currently, there two main configurations that can be used. One is to configure POP mailboxes and a POP server on the exchange host, and let all users use POP-enabled mail clients. The other possibility is to arrange for a periodic SMTP mail transfer from the exchange host to another, local SMTP exchange host which has been queuing all the outbound mail. Of course, since this solution does not allow full-time Internet access, it is not too preferred.

3.4.5. Secure Socket Layer Protocol

SSL is a layered protocol. It is located above a reliable transport layer, such as TCP, and below an application layer. However, SSL is not TCP\IP dependant and can run under other application protocols. It takes applications' data to be sent, fragments it into manageable blocks, compresses it (optionally), applies a Message Authentication Code (MAC) to the data, encrypts it and then transmits the result.

When data is received, the reverse process occurs: the data is decrypted, verified, decompressed, reassembled and then delivered to the application layer.

The SSL protocol consists of two major layers :

(1) SSL Handshake protocol and

(2) SSL Record protocol.

In addition, there are two other elements :

(1) SSL Change Cipher Spec Protocol, and

(2) SSL Alert Protocol.

The Handshake and the latter two components are responsible for the management of SSL communication setup and security parameters. The SSL Record Protocol is the lowest layer of the SSL protocol stack and all the other parts, including application data, operate on top of it. The Figure 3.5 shows the SSL protocol stack.

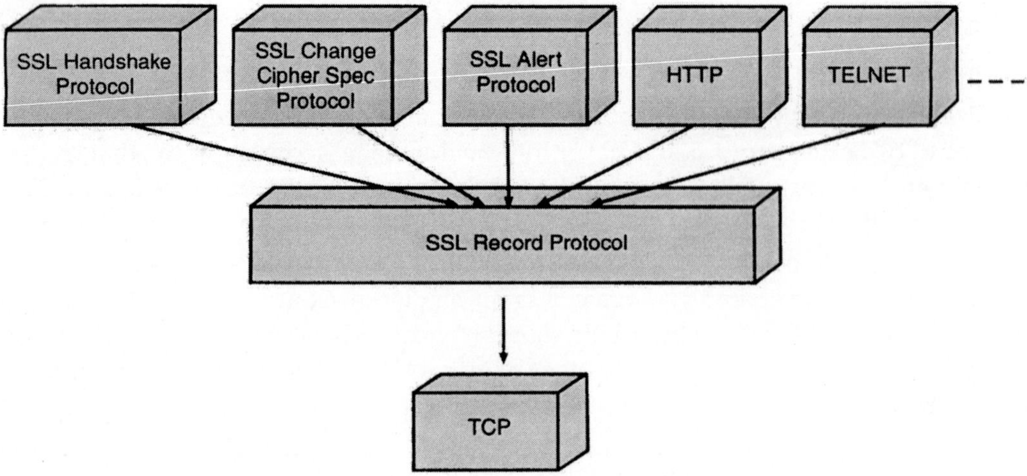

FIGURE 3.6. The SSL protocol slack.

3.5. WEB BROWSER

Web user access information from web servers, through **a client program** called **browser.**

Broadly speaking, a browser is responsible for the following tasks :

1. The first task is to accept a URL and retrieve the document specified by the URL.
2. The second task of browser is to interpret a HTML, document code, format it accordingly and finally render the document on the screen.
3. In this process, it may have to manage the rendering of various image formats, multimedia information, and links to other documents as well.

The most Famous Example of Browsers are Microsoft Internet Explorer and Netscape Navigator. Information regarding downloading the Netscape browser can be found at the site http://www.netscape.com. It can be download from the site or any of the listed mirror sites, and closest to your location. The internet explorer of Microsoft is popular browser, based on the Mossaic source code.

The overview of the architecture of the web browser is shown in Fig. 3.7.

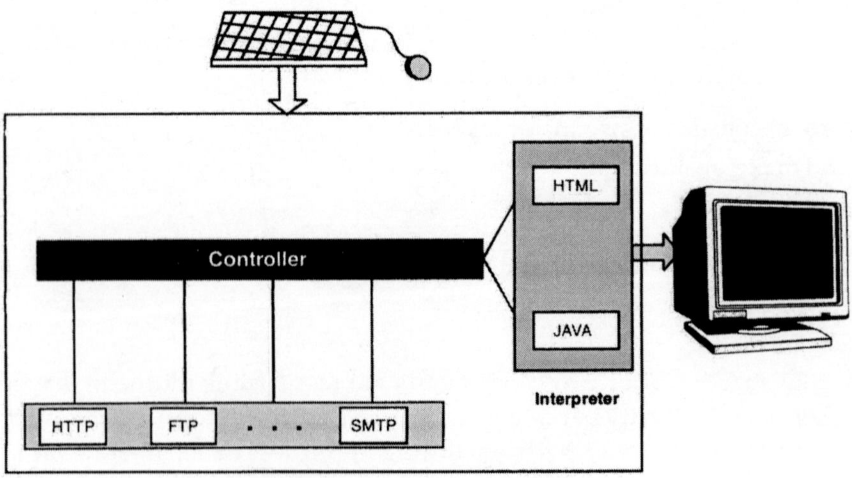

FIGURE 3.7. Browser architecture.

3.6. WEB SERVERS

A web server can mean two things—a computer on which a web site is hosted and a program that runs on such a computer. So the term web server refers to **both hardware and software**. We will look at each of these individually.

The Web Server Computer—The Hardware

A web site is a collection of web pages. And web pages are digital files, typically written using Hypertext Markup Language (HTML). For a web site to be available to everyone in the world at all times, it need to be stored or "hosted" on a computer that is connected to the internet .. Such a computer is known as a **web Server** .We can potentially host a web site on our home computer but this involves a lot of work and constant monitoring. It is easier to "buy" web hosting from a company because there are thousands that offer this service.

There are several requirements for a Server computer—it needs to be fast, have a large storage capacity hard disk and lots of RAM. But the most important is having a permanent internet address also known as an I.P. (Internet protocol) address. If the I.P. address changes, the web site would not be found and will appear offline—the browser will display a **cannot find web site** kind of error.

The Web Server Program—The Software

A web server program is software that runs on the web site hosting Server computer. Its main purpose is to serve web pages; which mean it waits for requests from web browsers (also known as **clients**) and responds by sending the required data back.

There are many web server programs available- check list of web servers. The most famous and popular of all web servers is **Apache** developed by the Apache Foundation. Not only is Apache free but it's also available for several operating systems including Windows, Macintosh and Linux/Unix.

3.6.1. Role of Web Server on Internet

Web servers—the computer or the program—have a vital role on the Internet. The Server machine hosts the web site while the server program helps deliver the web pages and their associated files like images and flash movies. The process of loading a web site/page in a web browser starts with the user either entering the URL in the address bar or clicking on a link. You should know that each web page has a unique address (or URL) on the internet; which means the same page cannot exist in two places. (If a copy does exist in another location, its address would be different from that of the original).

The browser now needs to send out a request for the web page. Behind the scenes, the URL of the requested web page is **resolved into an I.P. address**, which in English means, converted to an I.P. addresses—something that computers understand. The I.P. address points to the location of the web site host and the request is forwarded to Server computer and passed on to the server software. The server software now takes up and hunts for the requested web page on the hard disk. On finding the file, it sends back a response and the web page file to the browser which then starts displaying the page. A typical web page not only has text but also embedded multimedia elements like images and Flash animation. These "extra" files are separate from the actual web page and need to be sent one by one for the browser to display the web page correctly.

3.7. COMMON GATEWAY INTERFACE

The server manages HTML documents and delivers them to the browser, on request. These pages delivered by the server are usually HTML documents that have been prepared and placed on the server. For example, a list of items available in the company can be coded in a HTML document and placed on the server along with the inventory position. The change in the item inventory database need to be reflected on the product offering web page, served to the browser. Static HTML pages will require modification of the document every time the inventory database undergoes a change. A better solution would be to have a mechanism that generates the list of item document from the inventory database for each request of the document by any client/browser.

"The Common Gateway Interface (CGI) mechanism of HTTP enables servers to execute programs, obtain results and send the results to the requesting browser. The program executed by the by HTTP servers can be compiled using C, C++ programs or in scripting languages such as Unix shell (sh) or perl. These CGI programs also to referred as gateway programs/scripts, act as a script between the HTTP server and other programs such as DBMS."

Let us understand the action of the CGI by the Fig. 3.8.

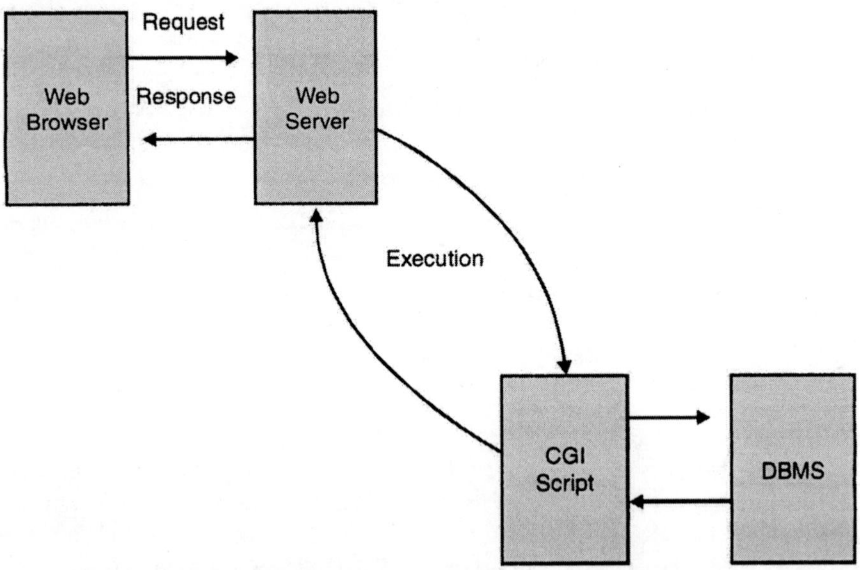

FIGURE 3.8. Interaction between browser and CGI program.

In above figure web browser send the request to web server, then the program related to this request is executed through CGI script and response back to client through web server.

3.7.1. CGI Scripts, CGI Programs or CGI Applications

The term we use will depend on the nature of the application that we are working on, although many people will tend to use and stick to a single term, there is a difference.

CGI Scripts - are interpreted by the server in real-time, and need to be interpreted each and every time they are run. For this reason they can be slower than a CGI program, but they are also easier to maintain as the code is accessible through a text editor.

CGI Programs - are compiled on the server they are running to help speed up the execution time and to hard code data paths for extra security. However, it can be difficult to compile programs on systems where access is limited, and the program will need to be recompiled for each platform we want to use it on.

CGI Applications - typically a CGI Application is a collection of CGI scripts or programs that work together to create a web-based interface for a program or database, although a single stand alone program or script can be described as an application if it does more than one type of server interaction.

3.7.2. Requirements

In general following software are required to execute the CGI :

(1) Web server (NCSA, Apache, IIS, Microsoft Personal Web server etc.)

(2) Compiler (C/C++) or Interpreter (Perl), PHP, ASP

(3) Web browser (NN, IE etc.)

3.8. HYPERTEXT MARKUP LANGUAGE (HTML)

HTML is a representation language for hypertext documents containing text, list boxes and graphics information. While HTML is great for displaying text among with limited graphics, it has many shortcoming that require frequent extensions to the language. In fact, one reason for Netscape's popularity has been the tradition of advancing the language by introducing new extensions. For instance, the most recent HTML addition is the Frame tags, which allows a window to be split into multiple independent panes. Each frame is an independent window-within-the-main window.

Let us see come of the tags of HTML for simple practical examples :

Beginning Tag	Ending Tag	Meaning
	Skeletal Tags	
<HTML>	</HTML>	Defines an HTML document
<HEAD>	</HEAD>	Defines the head of the document
<BODY>	</BODY>	Defines the body of the document
	Title and Header Tags	
<TITLE>	</TITLE>	Defines the title of the document
<Hn>	</Hn>	Defines the title of the document

Beginning Tag	Ending Tag	Meaning
	Text Formatting Tags	
<B>	</B>	Boldface
<I>	</I>	Italic
<U>	</U>	Underlined
_		Subscript
[	]	Superscript
	Data Flow Tag	
<CENTER>	</CENTER>	Centered
 	</BR>	Line break

Beginning Tag	Ending Tag	Meaning
	List Tags	
<OL>	</OL>	Ordered list
<UL>	</UL>	Unordered list
<LI>	</LI>	An item in a list
	Image Tag	
<IMG>		Defines an image
	Hyperlink Tag	
<A>	</A>	Defines an address (hyperlink)
	Executable Contents	
	</APPLET>	The document is an applet

This example shows how tags are used to let the browser format the appearance of the text.

```
<HTML>
        <HEAD>
            <TITLE> First Sample Document </TITLE>
        </HEAD>
        <BODY>
            <CENTER>
            <H1><B>  ATTENTION </B></H1>
            </CENTER>
```

You can get a copy of this document by :

```
            <UL>
             <LI> Writing to the publisher
             <LI> Ordering online
             <LI> Ordering through a bookstore
            </UL>
        </BODY>
</HTML>
```

This example shows how tags are used to import an image and insert it into the text.

```
<HTML>
        <HEAD>
            <TITLE> Second Sample Document </TITLE>
        </HEAD>
        <BODY>
            This is the picture of a book:
            <IMG SRC="Pictures/book1.gif" ALIGN=MIDDLE>
        </BODY>
</HTML>
```

This example shows how tags are used to make a hyperlink to another document.

```
<HTML>
        <HEAD>
            <TITLE> Third Sample Document </TITLE>
        </HEAD>
        <BODY>
    This is a wonderful product that can save you
  money and time.
        To get information about the producer, click on
          <A HREF="http://www.phony.producer">
          Producer </A>
        </BODY>
</HTML>
```

3.9. ELECTRONIC DATA INTERCHANGE

EDI developed in the 1960s as a mean of accelerating the movement of documents pertaining to shipments and transportation. Not until the mid-1980s, however, was the technique used in a wide range of industries – automotive, retail, transportation and international trade. Its use id growing and it is set to become the standard by which organization will communicate formally with each other in the word of electronic commerce.

Electronic commerce is often equated with EDI, so it is important to qualify that electronic commerce embraces EDI and much more. **In electronic commerce, EDI techniques are aimed at improving the interchange of information between trading partners, suppliers and customers by bringing down the boundaries that restrict how they interact and do business with each other.**

Now let us see various definition of the EDI as follows :

According to the Accredited Standard Committee for EDI of the American National Standards Institute. **"Electronic Data Interchange is the transmission, in a standard syntax, of unambiguous information of business or strategic significance of computers of independent organizations."**

According to the UN/EDIFACT *Training Guide* **: Electronic Data Interchange is the interchange of standard formatted data between computer application systems of trading partners with minimal manual invention.**

According to the *the European Model EDI Agreement* : **Electronic Data Interchange is the electronic transfer, from computer to computer of commercial and administrative data using an agreed standard to structure an EDI message.**

3.9.1. Building Blocks of EDI Systems : Layered Architecture

The general architecture of the EDI system consists of four layers: the application-conversion layer, standard message formats layer, the data transport layer and the interconnection layer as shown in Figure 3.9.

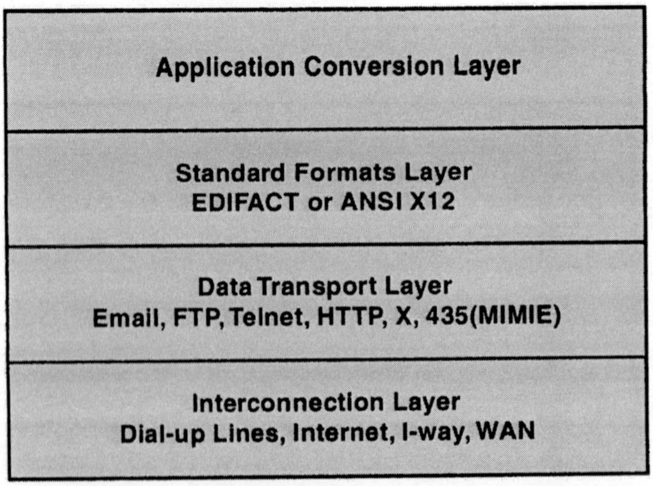

FIGURE 3.9. Layered architecture of EDI systems.

1. Application / Conversion Layer : The application layer consists of the actual business applications that are going to be connected through the EDI systems for exchange of electronic information. These applications may use their own electronic record formats and document formats for storing, retrieving and processing the information within each company's systems.

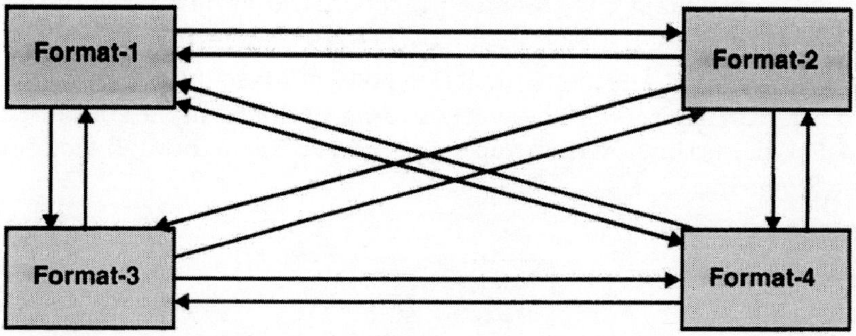

FIGURE 3.10. Converts between formats.

The problem of heterogeneity of formats can be better addressed using a common standard formats for documents/messages transferred within the EDI System. The internal processing systems continue to use the proprietary formats, but, for transmission over the wire, they adopt a common document/message format.

2. The Standard Formats Layer: The application layer of EDI systems rely on common agreed formats for operation. Thus, the second important and critical building block of the EDI systems is standards for business documents/forms. Since the sender and receiver in the EDI systems have to exchange business documents that can be interpreted by all parties, it has necessitated the development of standards in EDI. EDI forms standards are basically data standards in that they lay down the syntax and semantics of the data being exchanged.

The need for an industry-wide EDI standard was widely felt and this lead to the formation of a Standard Committee X12 under the auspices of American National Standards Institute (ANSI).

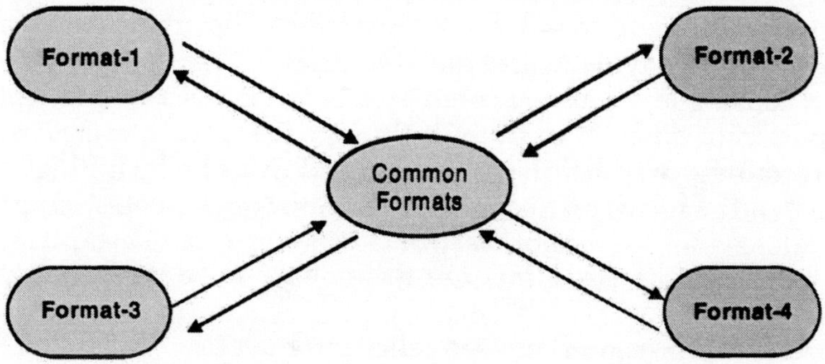

FIGURE 3.11. EDI common format.

3. Document Standards : The interconnection between trading partners only serves the purposing of exchanging information, but a document exchanged between two trading partners needs to be recognized and interpreted correctly by the correcting software systems running at various partner's computers.

Over a period of time, two major EDI standards have evolved. The first, commonly known as X12, was developed by the Accredited Standards X12 committee of American National Standards Institute (ANSI) and the second, the international standard, was developed by the United Nations EDI for Administration, Commerce and Trade (EDIFACT).

4. Data Transport Layer : The data transport layer consists of services that automata the task of electronic transfer of messages. The layer utilizes any of the available network transport services such as electronic mail; file transfer protocol, Telnet based remote connection and transfer, or even the Hypertext Transfer Protocol (HTTP) that derives the worldwide web.

5. Interconnection Layer : The interconnection layer refers to the network infrastructure that is used for the exchange of information between trading partners. In the simplest and most basic form it may consist of dial-up lines, where trading partners dial-up through modems to each other and connect to exchange messages.

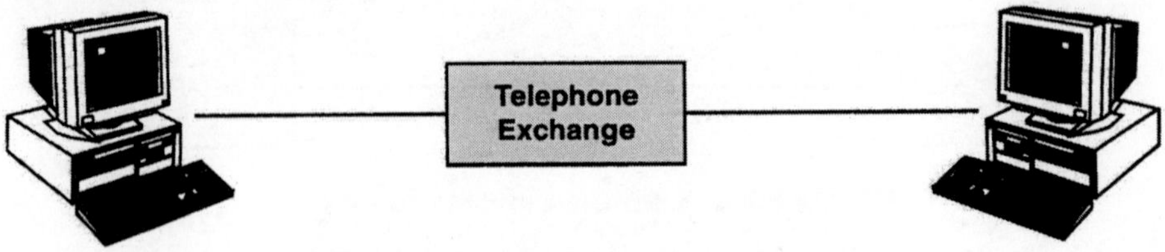

FIGURE 3.12. Dial-up Interconnection.

In case of the direct dial-up connections, partner computer have to be available for online connectivity and ready to receive the data at all times. Additionally, direct connections between partners have further problems as each partner has to establish a number of direct connections with all the partners. Also, for each partner, a variety of messages may originate, intended for other partners and of no relevance to a specific partner.

3.10. TRANSACTION

Transaction is exchange that occurs when one economic entity sells a product or service to another entity. A transaction takes place when a product or service is transferred across a technologically separable interface that links a consumer (client) with a producer (server).

To manage these transactions, electronic commerce also incorporates transaction management, which organizes, routes, processes and tracks transactions. Electronic commerce also includes consumers making electronic payments and funds transfer.

Another way of looking at electronic commerce is to view it as a production process that converts digital inputs into value-added outputs through the set of intermediaries. **For example, in the case of online trading, production process can add value by including more value-added processing (such as trend analysis) on the raw information (stock quotes) supplies to costumers.**

The business rationale for the use of electronic commerce can be explained by the simple question: Profit = Revenue – Costs. Firms use technology to either lower operating costs or increasing revenue.

Figure 3.13 represents transaction between seller and buyer.

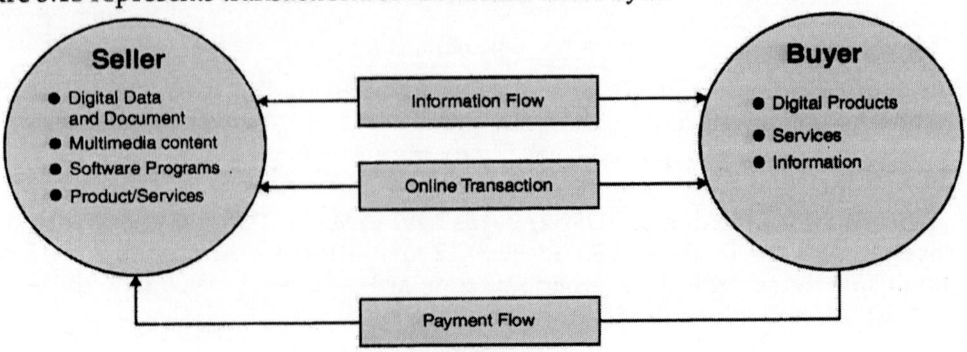

FIGURE 3.13.

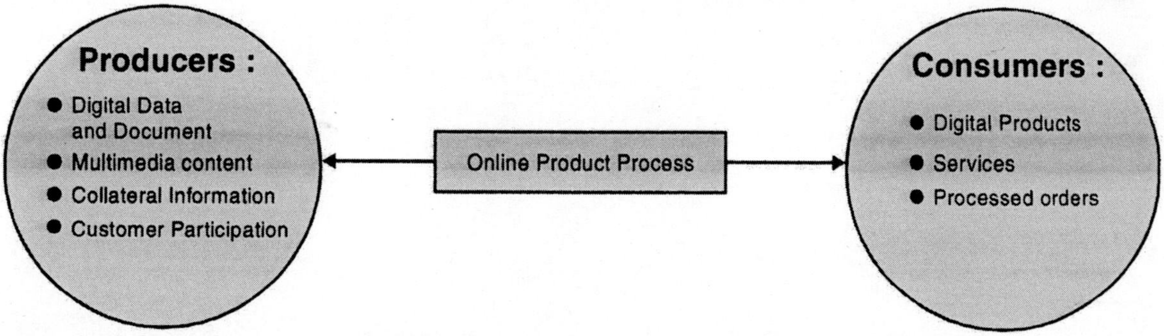

FIGURE 3.14. Product type transactions.

Figure 3.14 represents the transaction between producer and consumer

(1) Electronic commerce has the potential to increase revenue by creating new markets for old products, creating new information-based products, and establishing new service delivery channels to better serve and interact with costumer.

(2) The transaction management aspects of electronic commerce can also enable firms to reduce operating costs by enabling better coordination in the sales, production and distribution process (or better supply-chain management) and to consolidate operations and reduce overhead.

The goal of most electronic commerce research and its associated implementations is to reduce the "friction" in online transactions. Friction is often described in economics as transaction cost. Friction can arise from inefficient market structures linking buyers, sellers and intermediaries; inefficient organizational structures (operating units, business processes and workflows); and inefficient combinations of the technological activities required to make a transaction. Ultimately, the reduction of friction is online commerce will enable smoother transactions between buyers, intermediaries and sellers.

SUMMARY

In this chapter, we have seen the architectural requirement for the e-commerce. Now we are aware the role of internet and intranet in e-commerce what is use of web browser, web server and gateways of e-commerce on internet. We also discuss the protocol of internet that is TCP/IP. EDI is used to interchange the information between trading partner.

REVIEW QUESTIONS

1. What is internet and how it help in e-commerce ?
2. "E-commerce on the internet is risky". Comment on this statement.
3. Write the short notes on following :
   (*i*) URL (*ii*) DNS (*iii*) Web browser
   (*iv*) Web server (*v*) CGI
4. What kind of e-commerce architecture is required on the web to perform the e-commerce ?
5. What do you understand by basic architecture of web and what is the role of CGI in this architecture ?
6. What is EDI and discuss if layered architecture ?

❑❑❑

4

CHAPTER

NETWORK INFRASTRUCTURE FOR E-COMMERCE

4.1. INTERNET AND INTRANET BASED E-COMMERCE ISSUES

INSIDE THIS CHAPTER

4.1. Internet and Intranet Based E-Commerce Issues
4.2. Network for E-Commerce
4.3. Introduction to Broadband
4.4. Introduction to ATM
4.5. Introduction to ISDN
4.6. Frame Relay Introduction

It is clear that Internet and intranet are basic tools to implement the e-commerce. But as we know that Internet is free to every body so creates lot of problem for e-commerce implements. Let us see some of them as follows :

1. Security

While shopping on the Internet, most people typically do not think about what is happening in the background. Web shopping is generally very easy. We click on a related site, go into that site, buy the required merchandise by adding it to our cart, enter our credit card details and then expect delivery within a couple of days. This entire process looks very simple but a developer or businessmen knows exactly how many hurdles need to be jumped to complete the order.

2. Privacy

Customer information has to pass through several hands so security and privacy of the information are a major concern. The safety and security of a customer's personal information lies within the hands of the business. Therefore businesses have to give the customer first their guarantee, and second peace of mind, that the information passed over is of no risk to any invading eyes.

Many people are not willing to disclose their personal information on the Web. It is up to individuals to decide how much personal information they are willing to disclose and how it might be used. Interestingly, one survey found that many people who disclose personal information do so in hope of financial benefit.

3. Copyright

Many attempts have been made to address the issues related to copyrights on digital content. E-commerce has a tremendous impact on copyright and related issues, and the scope of copyrights is affecting how e-commerce evolves. It is essential that legal rules are set and applied appropriately to ensure that digital technology does not undermine the basic doctrine of copyright and related rights. From one perspective, the Internet has been described as "the world's biggest copy machine".

4. Trademark

Generally, an individual, a company, or any sort of legal entity can own a trademark. When someone else tries to use that trademark (e.g., your distinctive name or logo) without authorization, it could be considered an illegal dilution of the distinctive trademark. If someone uses a trademark in such a way as to dilute the distinctive quality of the mark or trade on the owner's reputation, the trademark owner may seek damages.

Some Web-based applications have enabled large-scale exploitation of music samples and audio formats. Software that is available free of cost on the Net allows the transfer of songs and videos without the authorization of rights holders. Moreover, CD burners and portable MP3 players allow copyright violations to occur rather easily.

5. Online Terms, Conditions, Policies and Laws

At the moment, most online privacy policies are produced by private businesses for individual companies. Governments are developing legislation to support and strengthen the privacy protection measures of many businesses. These initiatives are aimed at regulating the storage, use and disclosure by businesses of personal information.

Privacy legislation is designed to protect a person's personal information. The privacy laws of their host country affect overseas companies. Every organization should be very careful while applying terms and conditions for the electronic transaction for Internet users. Privacy and security policies not only reflect the organizations practice but also the rules and regulations for doing business with the company. Major issues regarding the legalization of electronic transactions include the following :

(*a*) Ensure proper online contracts.

(*b*) Record retention obligations.

(*c*) Original documentation, in terms of TAX and VAT requirements.

(*d*) Import/export regulations.

(*e*) Exchange control regulation.

(*f*) Foreign data protection law.

6. Legislation Dilemma

Electronic transactions separate e-business from traditional types of businesses. When a transaction takes place, Who has jurisdiction ? Who has the authority to apply law over the transaction ?

For example, if you buy a laptop in your local computer store, you know your legal rights. If the computer does not work when you take it home, and the store refuses to settle up, then you can probably take the dispute to your local small claims court. But if you buy the same computer online, from a vendor on the other side of the world, perhaps through a dealer based in yet a third country, then your rights are a lot less clear. Which country's protection laws apply: yours, those in the vendor's home country, or those of the intermediary ? Without knowing which particular set of laws apply, it's impossible to know whom to sue. "Small claims courts don't work in cyberspace," according to Ron Presser of the American Bar Association.

4.2. NETWORK FOR E-COMMERCE

In any country successful implementation of e-commerce depends on the infrastructure available. Network for commerce can be discussed at following three levels:

(*i*) Corporate Networks

(*ii*) Commercial Networks

(*iii*) Internet

The network can be lay down at corporate level in the form of LAN and WAN to manage the internal commerce of the organizations.

Commercial activities can be managed through the common networks shared by corporate, consumers and others involves. The common network may be LAN , WAN, MAN or ultimately internet. Through internet easy and cheap communication is possible any where and with any one.

4.2.1. Introduction to I-Way

Electronic commerce needs a network infrastructure *to transport the content – data, audio, visual, text, animation and so on.* This network infrastructure is provided by what is known as the *I-way or business super highway. The information super highway may be defined as* ***"a high capacity, electronic pipeline to a consumer or business premises that is capable of simultaneously supporting a large number of electronic commerce applications and providing interactive connectivity between users and services and between users and other users".***

The traditional communication infrastructure has evolved separately for voice and data network, voice network relying on circuit switching and data networks using packet switching techniques. The commercial user requires, voice, data and video conferencing services. The development of integrated electronic commerce applications in manufacturing, health, education, banking and insurance and other industries is paying the way for a network infrastructure providing the support to various types of information.

4.2.2. Requirement of I-Way

The success of E-Commerce based business depends on the information flow and to make information flow smooth an capable I-WAY (information super highway) is required .

The success or failure of any creative, product, or services is a key driver of market forces. The understanding of market drivers of I-way is important because e-commerce applications are dependent on the underlying I-way.

The I-way will provide the supports :

(*i*) Consumers, end users or business consuming and paying for information product/services.

(*ii*) Users who become information publishers by setting up online servers.

(*iii*) Value-added information providers, including third party brokers and other intermediaries, as well as originators of services who add value by packing or building on service provided by others.

(*iv*) Information service providers, who are commercial, government, or private providers, or publishers of information goods and services.

Users and enterprise play *multiple roles as consumer* and *producers* of *information*. These roles are not fixed but can be combined in various ways such that enterprise can, for example, concurrently be information consumers and service providers. Various types of user roles provides an indication of the market structure and could explain why many companies are merging or realigning themselves.

4.2.3. Components of the I-Way

The I-way consists of various components, which can be broadly categorized as

1. Network Access Equipment

To access any network some specific equipments are required at the costumer end and enables the consumer to access the network. It consists of the hardware such as computers, modems, routers, switches for computer networks, set-top boxes for television networks and software platform such as browsers and operating systems.

2. Access Media

A suitable bandwidth and transmission media is required for the successful implementation of e-commerce .Any nation should have good communication backbone to implement the e-commerce.

Access media can be divided in to four categories :

- Telecom Media,
- Cable TV
- Wireless Media
- Internet, Intranet and Extranet (Online Media)

The function of the access providers is to link the users and the e-commerce application providers such as the telecom networks, satellite networks and the Internet.

3. Global Information Distribution Network

Now days e-commerce activities are not limited to any country only but now world becomes a global village. So a proper network infrastructure is required to connect the people and business across the world. These types of distribution network may include following type of networks:

(*a*) Long Distance Networks : **Long distance telephone connectivity is provided through cable (co-axial or fiber) by the inter exchange carriers. Long distance cellular networks are using the wireless technologies to connect the consumers worldwide.**

(*b*) Satellite Networks : **Satellite network plays a vital role in the telecommunication industry. They have advantages over the terrestrial networks in that :**

(*i*) They are accessible from any point on the global

(*ii*) They can provide broadband digital services to many points without the cost of acquiring wire/cable installation.

(*iii*) They can add receiving and sending sites without significant additional costs.

Communications satellites form a crucial part of the global communication infrastructure, providing a wide range of services including broadcast video and overseas telephone links.

(c) Internet : The Internet forms a well-known component of the global information distribution network. It targets a wide range of e-commerce applications such as video on demand, home shopping, e-mail, EDI, information publishing, information retrieval, video conferencing and many more.

All the components of I-way together provide a network infrastructure for the E-commerce activities. This requires the use of common standards and installing gateways between various networks. A final requirement is the hardware and software to move huge amounts of data effortlessly over the complex networks. Let us see following figure :

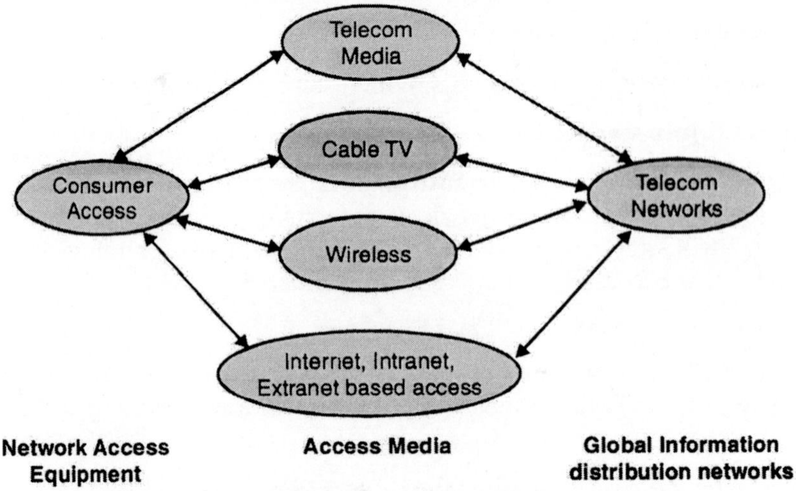

FIGURE 4.1. Components of the information super highway.

It is clear form Figure 4.1 that consumers can interact all over the world through various types of media available on communication networks.

4.3. INTRODUCTION TO BROADBAND

High-speed Internet access or "broadband" allows users to access the Internet and Internet-related services at significantly higher speeds than those available through "dial-up" Internet access services. The Federal Communications Commission (FCC) generally defines broadband service as data transmission speeds exceeding 200 kilobits per second (Kbps), or 200,000 bits per second, in at least one direction: downstream (from the Internet to your computer) or upstream (from your computer to the Internet).

4.3.1. Working of Broadband

Broadband allows users to access information via the Internet using one of several high-speed transmission technologies. Transmission is digital, meaning that text, images, and sound are all transmitted as "bits" of data. The transmission technologies that make broadband access possible move these bits much more quickly than traditional telephone or wireless connections, including traditional dial-up Internet access.

Once we have a broadband connection to our home or business, devices such as computers can be attached to this broadband connection by existing electrical or telephone wiring, coaxial cable, or wirelessly.

4.3.2. Advantages of Broadband

Broadband allows us to take advantage of new services not available with a dial-up Internet connection. One such service is Voice over Internet Protocol (VoIP), an alternative to traditional voice telephone service that may be less costly for us depending on our calling patterns. Some VoIP services only allow us to call other people using the same service, but others allow us to call anyone who has a telephone number – including local, long distance, mobile, and international numbers.

Broadband permits new developments in telemedicine, where patients in rural areas can confer online with medical specialists in more urban areas.

Broadband helps us efficiently access and use many reference and cultural resources, such as library and museum data bases and collections. We also need broadband to best take advantage of many distance learning opportunities, like online college or university courses, and continuing or senior education programs. Broadband is an important tool for expanding educational and economic opportunities for consumers in remote locations.

In addition to these new services, broadband allows us to use existing services such as online shopping and web surfing more quickly and efficiently. Downloading and viewing videos and photos on our computer are much faster and easier. With broadband we can access the Internet by turning on our computer without needing to dial-up your Internet Service Provides (ISP) over a telephone line, which permits us to use the Internet without typing up our telephone line. As of the end of 2004, 35.3 million residential and small business subscribers had opted for broadband connections.

4.3.3. Types of Broadband

Broadband includes several high-speed transmission technologies such as :

- Digital Subscriber Line (DSL)
- Cable Modem
- Fiber
- Wireless
- Satellite
- Broadband over Powerline (BPL)

1. Digital Subscriber Line (DSL)

DSL is a wireline transmission technology that transmits data faster over traditional copper telephone lines already installed to homes and businesses. DSL-based broadband provides transmission speeds ranging from several hundred Kbps to millions of bits per second (Mbps). The availability and speed of your DSL service may depend on the distance from your home or business to the closest telephone company facility.

The following are types of DSL transmission technologies :

(*a*) **Asymmetrical Digital Subscriber Line (ADSL)** used primarily by residential customers, such as Internet surfers, who receive a lot of data but do not send much. ADSL typically provides faster speed in the downstream direction than the upstream direction. ADSL allows faster downstream data transmission over the same line used to provide voice service, without disrupting regular telephone calls on that line.

(*b*) **Symmetrical Digital Subscriber Line (SDSL)** used typically by businesses for services such as video conferencing. Speed of downstream and upstream traffic is equal.

Faster forms of DSL typically available to businesses include :

- High-data-rate Digital Subscriber Line (HDSL); and
- Very High-data-rate Digital Subscriber Line (VDSL).

2. Cable Modem

Cable modem service enables cable operators to provide broadband using the same coaxial cables that deliver pictures and sound to your TV set.

Most cable modems are external devices that have two connections, one to the cable wall outlet and the other to a computer. They provide transmission speeds of 1.5 Mbps or more.

3. Fiber

Fiber, or fiber optics, is the newest technology available for providing broadband. Fiber optic technology converts electrical signals carrying data to light and sends the light through transparent glass fibers about the diameter of a human hair. Fiber transmits data at speeds far exceeding current DSL or cable modem speeds, typically by tens or even hundreds of Mbps. However, the actual speed you experience will vary depending upon a variety of factors, such as how close to your computer the service provider brings the fiber, and how the service provider configures the service, including the amount of bandwidth used. The same fiber providing your broadband can also simultaneously deliver voice (VoIP) and video services, including video-on-demand.

4. Wireless

Wireless broadband connects a home or business to the Internet using a radio link between the customer's location and the service provider's facility. Wireless broadband can be mobile or fixed.

Wireless technologies using longer range directional equipment provide broadband service in remote or sparsely populated areas where DSL or cable modem service would be costly to provide. Speeds are generally comparable to DSL and cable modem. An external antenna is usually required. With newer services now being deployed (WiMax), a small antenna located inside a home near a window is usually adequate and higher speeds are possible.

Fixed wireless broadband service is becoming more and more widely available at airports, city parks, bookstores, and other public locations called "hotspots." Hotspots generally use a short-range technology that provides speeds up to 54 Mbps. Wireless fidelity (Wi-Fi) technology is also often used in conjunction with DSL or cable modem service to connect devices within a home or business to the Internet via a broadband connection.

Mobile wireless broadband services (3G) are also becoming available from mobile telephone service providers and others. These services are generally appropriate for highly-mobile customers and require a special PC card with a built in antenna that plugs into a user's laptop computer. Generally, they provide lower speeds, in the range of several hundred Kbps.

5. Satellite

Just as satellites orbiting the earth provide necessary links for telephone and television service, they can also provide links for broadband. Satellite broadband is another form of wireless broadband, also useful for serving remote populated areas.

Downstream and upstream speeds for satellite broadband depend on several factors, including the provider and service package purchased, the consumer's line of sight to the orbiting satellite, and the weather. Typically a consumer can expect to receive (download) at a speed of about 500 Kbps and send (upload) at a speed of about 80 Kbps. These speeds may be slower than DSL and cable modem, but download speed is about 10 times faster than download speed with dial-up Internet access. Service can be disrupted in extreme weather conditions.

Obtaining satellite broadband can be more costly and involved than obtaining DSL or cable modem. A user must have :

- a two or three foot dish or base station – the most costly item;
- a satellite Internet modem; and
- a clear line of sight to the provider's satellite.

6. Broadband over Powerline (BPL)

BPL is the delivery of broadband over the existing low and medium voltage electric power distribution network. BPL speeds are comparable to DSL and cable modem speeds. BPL can be provided to homes using existing electrical connections and outlets.

BPL is an emerging technology, currently available in very limited areas. It has significant potential because power lines are installed virtually everywhere, alleviating the need to build new broadband facilities to every customer.

4.4. INTRODUCTION TO ATM

Asynchronous Transfer Mode (ATM) technology will play a central role in the evolution of current workgroup, campus and enterprise networks. ATM delivers important advantages over existing LAN and WAN technologies, including the promise of scalable bandwidths at unprecedented price and performance points and Quality of Service (QoS) guarantees, which facilitate new classes of applications such as multimedia.

These benefits, however, come at a price. Contrary to common misconceptions, ATM is a very complex technology, perhaps the most complex ever developed by the networking industry. While the structure of ATM cells and cell switching do facilitate the development of hardware intensive, high performance ATM switches, the deployment of ATM *networks* requires the overlay of a highly complex, software intensive, protocol infrastructure. This infrastructure is required to both allow individual ATM switches to be linked into a network, and for such networks to internetwork with the vast installed base of existing local and wide area networks.

4.5. INTRODUCTION TO ISDN

Integrated Services Digital Network is a digital telephone service that provides fast, accurate data transmission over existing copper telephone wiring.

ISDN is based on a number of fundamental building blocks. First, there are two types of ISDN "channels" or communication paths:

(*i*) **B-channel :** The Bearer ("B") channel is a 64 kbps channel which can be used for voice, video, data, or multimedia calls. B-channels can be aggregated together for even higher bandwidth applications.

(*ii*) **D-channel :** The Delta ("D") channel can be either a 16 kbps or 64 kbps channel used primarily for communications (or "signaling") between switching equipment in the ISDN network and the ISDN equipment at your site.

These ISDN channels are delivered to the user in one of two pre-defined configurations :

(*a*) **Basic Rate Interface (BRI) :** BRI is the ISDN service most people use to connect to the Internet. An ISDN BRI connection supports two 64 kbps B-channels and one 16 kbps D-channel over a standard phone line. BRI is often called "2B+D" referring to its two B-channels and one D-channel. The D-channel on a BRI line can even support low-speed (9.6 kbps) X.25 data, however, this is not a very popular application in the United States.

(*b*) **Primary Rate Interface (PRI) :** ISDN PRI service is used primarily by large organizations with intensive communications needs. An ISDN PRI connection supports 23, 64 kbps B-channels and one 64 kbps D-channel (or 23B+D) over a high speed DS1 (or T-1) circuit. The European PRI configuration is slightly different, supporting 30B+D.

4.5.1. Use of ISDN

ISDN offers the speed and quality that previously was only available to people who bought expensive, point-to-point digital leased lines. Combined with its flexibility as a dial-up service, ISDN has become the service of choice for many communications applications. Popular ISDN applications include :

- Internet access.
- Telecommuting/remote access to corporate computing.
- Video conferencing.
- Small and home office data networking.

4.5.2. Why ISDN to Access the Internet

More and more people are discovering that ISDN is the right Internet answer.

As the Internet becomes more and more information-intensive with graphics, sound, video and multimedia, our ability to take advantage of these new resources depends on the speed of our Internet connection. Can our existing connection handle these large files quickly and cleanly ? Does it take forever to download files ? Are our downloads frequently aborted because of transmission errors ?

With ISDN, our Internet access is :

- **Even Faster :** By combining our two B-channels we have access to up to 128 kbps — more than four times as fast as a 28.8 kbps modem on a standard phone line. And ISDN's digital technology assures us the cleanest connection to the Internet so we won't be slowed down by re-transmissions because of old analog technology.
- **More Efficient and Economical :** ISDN brings increased capabilities, reduced costs and improved productivity to organizations both large and small. When we're looking for something on the Internet, we can get there faster. We can be more productive because we aren't waiting as long to get to that next website or download that large file.

4.6. FRAME RELAY INTRODUCTION

Frame Relay is a high-performance WAN protocol that operates at the physical and data link layers of the OSI reference model. Frame Relay originally was designed for use across Integrated Services Digital Network (ISDN) interfaces. Today, it is used over a variety of other network interfaces as well. This chapter focuses on Frame Relay's specifications and applications in the context of WAN services.

Frame Relay is an example of a packet-switched technology. Packet-switched networks enable end stations to dynamically share the network medium and the available bandwidth. The following two techniques are used in packet-switching technology :

(*i*) Variable-length packets

(*ii*) Statistical multiplexing

Variable-length packets are used for more efficient and flexible data transfers. These packets are switched between the various segments in the network until the destination is reached.

Statistical multiplexing techniques control network access in a packet-switched network. The advantage of this technique is that it accommodates more flexibility and more efficient use of bandwidth. Most of today's popular LANs, such as Ethernet and Token Ring, are packet-switched networks.

Frame Relay often is described as a streamlined version of X.25, offering fewer of the robust capabilities, such as windowing and retransmission of last data that are offered in X.25. This is because Frame Relay typically operates over WAN facilities that offer more reliable connection services and a higher degree of reliability than the facilities available during the late 1970s and early 1980s that served as the common platforms for X.25 WANs. As mentioned earlier, Frame Relay is strictly a Layer 2 protocol suite, whereas X.25 provides services at Layer 3 (the network layer) as well. This enables Frame Relay to offer higher performance and greater transmission efficiency than X.25, and makes Frame Relay suitable for current WAN applications, such as LAN interconnection.

4.6.1. Frame Relay Standardization

Initial proposals for the standardization of Frame Relay were presented to the Consultative Committee on International Telephone and Telegraph (CCITT) in 1984. Because of lack of interoperability and lack of complete standardization, however, Frame Relay did not experience significant deployment during the late 1980s.

A major development in Frame Relay's history occurred in 1990 when Cisco, Digital Equipment Corporation (DEC), Northern Telecom, and StrataCom formed a consortium to focus on Frame Relay technology development. This consortium developed a specification that conformed to the basic Frame Relay protocol that was being discussed in CCITT, but it extended the protocol with features that provide additional capabilities for complex internetworking environments. These Frame Relay extensions are referred to collectively as the Local Management Interface (LMI).

Since the consortium's specification was developed and published, many vendors have announced their support of this extended Frame Relay definition. ANSI and CCITT have subsequently standardized their own variations of the original LMI specification, and these standardized specifications now are more commonly used than the original version.

Internationally, Frame Relay was standardized by the International Telecommunication Union—Telecommunications Standards Section (ITU-T). In the United States, Frame Relay is an American National Standards Institute (ANSI) standard.

4.6.2. Frame Relay Devices

Devices attached to a Frame Relay WAN fall into the following two general categories:

(*i*) Data terminal equipment (DTE)

(*ii*) Data circuit-terminating equipment (DCE)

DTEs generally are considered to be terminating equipment for a specific network and typically are located on the premises of a customer. In fact, they may be owned by the customer. Examples of DTE devices are terminals, personal computers, routers, and bridges.

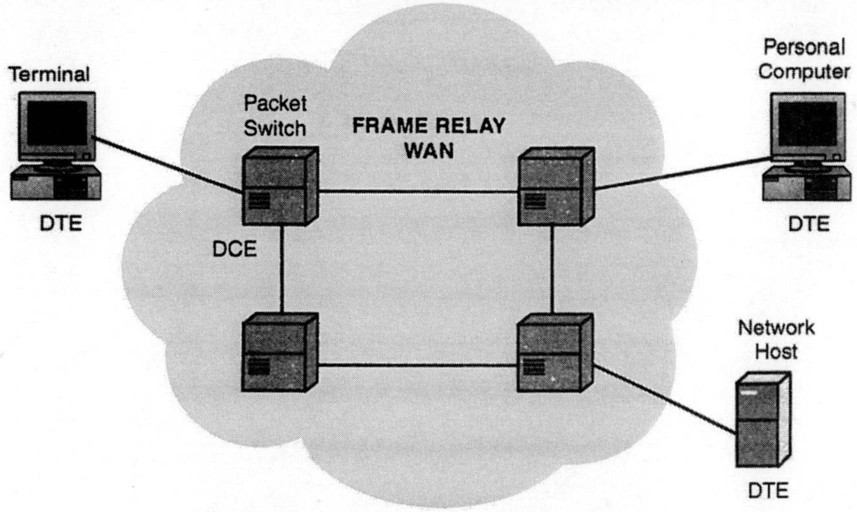

FIGURE 4.2. Frame relay devices in WAN.

DCEs are carrier-owned internetworking devices. The purpose of DCE equipment is to provide clocking and switching services in a network, which are the devices that actually transmit data through the WAN. In most cases, these are packet switches. Figure 4.2 shows the relationship between the two categories of devices.

The connection between a DTE device and a DCE device consists of both a physical layer component and a link layer component. The physical component defines the mechanical, electrical, functional, and procedural specifications for the connection between the devices. One of the most commonly used physical layer interface specifications is the recommended standard (RS)-232 specification. The link layer component defines the protocol that establishes the connection between the DTE device, such as a router, and the DCE device, such as a switch.

SUMMARY

In this chapter we learn that in any country successful implementation of e-commerce depends on the infrastructure available. Network for commerce can be discussed at levels like corporate networks, commercial networks and Internet. The success of E-Commerce based business depends on the information flow and to make information flow smooth an capable I-WAY (information super highway) is required. We connect our self with I-Way through various means like fiber optics, satellite and ISDN.

REVIEW QUESTIONS

1. Now a days almost all e-commerce is performed through Internet and intranet, discuss the related issues for such e-commerce.
2. What do you understand by I-Way and what is the need of it ?
3. Discuss various component of Information super high way.
4. What do you understand by broadband ? Discuss the working of it.
5. What are various types of broadband ?
6. What do you understand by ATM ?
7. Why should we select ISDN to access the Internet ?
8. Write the short notes on following :
   (*a*) Frame Relay
   (*b*) ISDN
   (*c*) Satellite
   (*d*) DSL
   (*e*) Fiber Optics
   (*f*) Cable Modem
   (*g*) Wireless

❑❑❑

5

CHAPTER

MOBILE COMMERCE

INSIDE THIS CHAPTER

5.1. Introduction to Mobile Commerce
5.2. M-Commerce Versus E-Commerce
5.3. Driving Force for M-Commerce
5.4. Limiting Factors
5.5. Mobile Customer and Applications
5.6. Applications of M-Commarce
5.7. Mobile Computing
5.8. Introduction to WAP (Wireless Application Protocol)

5.1. INTRODUCTION TO MOBILE COMMERCE

There is no precise definition for mobile commerce or mobile e-commerce. Every analyst, vendor and trade magazine appears to have a slightly different interpretation of the term. In fact, there appears to be a consensus now on the meaning of its big brother "e-commerce or e-business" which came first. Perhaps, in another six months or a year, we would come to a similar consensus on M-commerce. Meanwhile, here are a few sample definitions :

"M-Commerce is the use of mobile devices to communicate, inform transact and entertain using text and data via a connection to public and private networks." **(Lehman Brothers)**

"The core of mobile e-commerce is the use of a terminal (telephone, PDA, PC device, or custom terminal) and public mobile network (necessary but not sufficient) to access information and conduct transactions that result in the transfer of value in exchange for information, services or goods." **(Ovum)**

"Business-to-consumer transactions conducted from a mobile device." **(J.P. Morgan)**

"E-Commerce over mobile devices." **(Robinson-Humphreys)**

"Mobile Commerce refers to any transaction with monetary value that is conducted via a mobile telecommunications network." **(Durlacher)**

"The use of mobile handheld devices to communicate, interact via an always-on high-speed connection to the Internet." **(Forrester)**

"The use of wireless technologies to provide convenient personalized and location-based services to your customers, employees and partners." **(Mobilocity)**

Typical examples of M-commerce are :

- Purchasing airline tickets
- Purchasing movie tickets
- Restaurant booking and reservation
- Hotel booking and reservation

5.2. M-COMMERCE VERSUS E-COMMERCE

Frequently M-commerce is represented as a "subset of all e-commerce" thus implying that any e-commerce site could and should be made available from a wireless device. *We believe that such* conclusions are misleading. M-commerce should be recognized as a unique business opportunity with its own unique characteristics and functions, not just an extension of an organization's Internet-based e-commerce channel. Of course there are similarities between e-commerce and M-commerce from being able to purchase a product or service in a "virtual" vs. a build and mortar environment. The following table will clearly differentiate between E-Commerce and M-Commerce.

Technology	E-Commerce	M-Commerce
Device	PC	Smartphones, pagers, PDAs,
Operating System	Windows, Unix, Linux	PalmOS, Pocket PC, proprietary platforms.
Presentation Standards	HTML	HTML, WML, HDML, i-Mode
Browser	Microsoft Explorer, Netscape	Nokia browser, MS Mobile Explorer and other microbrowsers
Bearer Networks	TCP/IP & Fixed Wireline Internet	GSM, GSM/GPRS, TDMA, CDMA, CDPD, paging networks

5.3. DRIVING FORCE FOR M-COMMERCE

The fast growth is the main reason behind the M-Commerce, Let us discuss some factors which are driving force of M-commerce as follows :

(*i*) Exponential growth of consumer interest and adoption of the Internet and e-commerce.

(*ii*) Tremendous growth in mobile telephony; however, voice has become a commodity and will no longer fuel revenue growth for operators.

(*iii*) Development of real-time transfer of data over 2.5G and 3G networks will enable faster data transmission and 'always-on' connectivity.

(*iv*) The evolution of the handheld devices incorporating WAP and now GPRS.

(*v*) Mobile e-commerce market is worth $3.5 billion in 2000 and will grow to over $200 billion by 2005 (Ovum).

(*vi*) Cost of entry into mobile e-commerce is low for most entrants; for example, a bank can implement a sophisticated M-banking solution in under six months for around $1 million.

(*vii*) The unique features of the mobile device such as its compactness for convenience and personalized functions; subsequently, people have become quite attached to their devices.

M-Commerce is attracting players from all economic sectors from technology, finance, retail, media, all anticipating that M-commerce will increase customer acquisition and retention and generate new revenue opportunities.

5.4. LIMITING FACTORS

There are following two main limiting factors are there for M-commerce:

1. Lack of Standards

With a host of device operating systems and platforms, middleware solutions and networks, make application development for the wireless Internet a difficult task, versus the level operating environment of the wired Web.

2. Device Constraints

Proper device is necessary to implement M-commerce but still devices have their limitations like :

(*i*) Weak processors.
(*ii*) Limited memory.
(*iii*) Tiny screens, poor resolutions.
(*iv*) Poor data entry.

5.5. MOBILE CUSTOMER AND APPLICATIONS

"Customers will use wireless devices to access the Internet only if they have a good reason to do so". Fundamental attributes common to the mobile customer are as follows :

1. Expectations

Basically M-commerce users have some unrealistic expectations. Let us see some common expectations of consumers :

(*i*) Proper education is required.
(*ii*) Don't Set Unrealistic Expectations: Limit Hyped-Up possibilities, such as, Wireless Web – misleading – one can not push a Web service onto the cellphone.
(*iii*) Newness of the Wireless Internet attracting media and marketing hype, promoting unrealistic expectations among consumers.
(*iv*) Offering customers something new that could not do without a wireless device.

2. Voice is Still the Killer Application

(*i*) Most customers use wireless phones for voice phone calls.
(*ii*) Wireless data features are much less popular than wireless voice.

3. Personalization

Basically people start wireless devices with the personalized services/customized information and for the services via the Internet :

(*i*) Wireless devices as a personal companion.
(*ii*) Expect companies to offer information relevant to their needs and interests and presented in a simplified manner.
(*iii*) Seeking an easy and simplified M-commerce experience.
(*iv*) People expect only relevant advertising.
(*v*) Personalized billing/payment functionality are main attraction of M-commerce.

4. Dependency on location and Time

(*i*) Understand the customer's activity schedule, for instance, services accessed during working hours will differ from those on during the evening or on weekends.
(*ii*) Device used will determine mode of access.

5. Consumer Behaviour

(*i*) Patience will be demanded of players, as consumer behavior requires both time and effort to change.

(*ii*) Evolution of the industry from networks to handsets will require players to educate their customers.

5.6. APPLICATIONS OF M-COMMERCE

According to Ovum's research, there is a lot of uncertainty about which mobile commerce applications will be successful and make money. The research/consulting firm classified M-commerce applications into three categories.

1. Goods : **There can be two type of domain like** business-to-business and business to customer. **For example a vendor can sail to customer or to another vendor.**

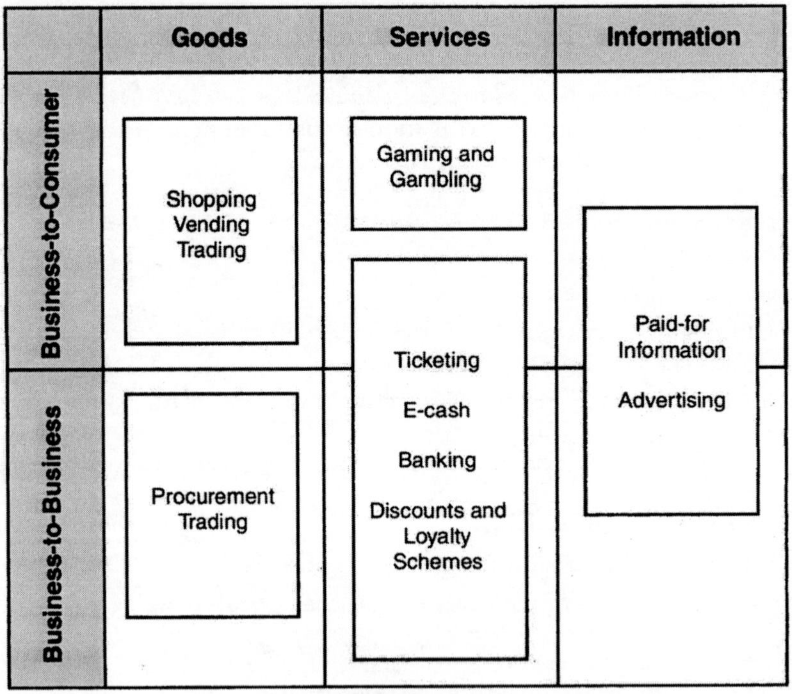

FIGURE 5.1.

2. Service : **Again service can be** business to business or customer to business. **For example a company can provide service to directly to customers or to another company.**

3. Information : **Information service can be paid to customers or to others companies. The second aspect is advertising.**

In Figure 5.1 the M-commerce application in business-to-consumer and business-to-business can be easily understood. M-Commerce can provide business comfort to organizations and also can be very help full to consumers.

5.7. MOBILE COMPUTING

A technology that allows transmission of data, via a computer, without having to be connected to a fixed physical link is said to be Mobile Computing.

Mobile voice communication is widely established throughout the world and has had a very rapid increase in the number of subscribers to the various cellular networks over the last few years. An extension of this technology is the ability to send and receive data across these cellular networks. This is the principle of mobile computing.

Mobile data communication has become a very important and rapidly evolving technology as it allows users to transmit data from remote locations to other remote or fixed locations. This proves to be the solution to the biggest problem of business people on the move.

5.7.1. Existing Cellular Network Architecture

Mobile telephony took off with the introduction of cellular technology which allowed the efficient utilisation of frequencies enabling the connection of a large number of users. During the 1980's analogue technology was used. Among the most well known systems were the NMT900 and 450 (Nordic Mobile Telephone) and the AMPS (Advanced Mobile Phone Service). In the 1990's the digital cellular technology was introduced with GSM (Global System Mobile) being the most widely accepted system around the world. Other such systems are the DCS1800 (Digital Communication System) and the PCS1900 (Personal Communication System).

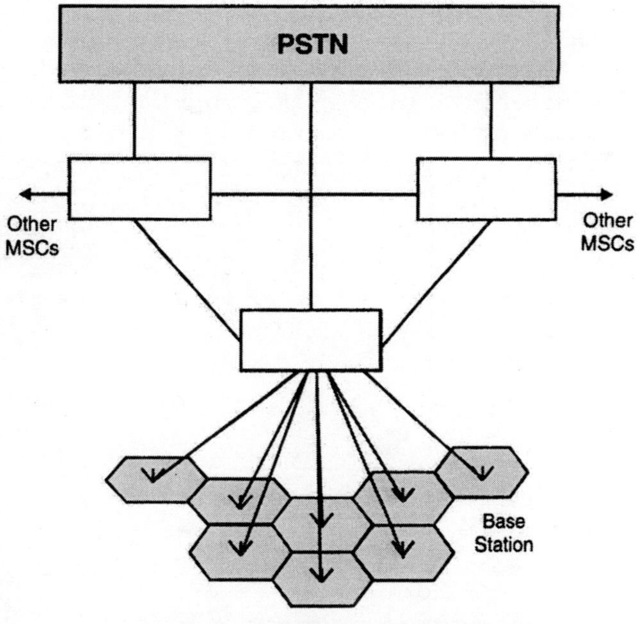

FIGURE 5.2. Mobile switching centre.

A cellular network consists of mobile units linked together to switching equipment, which interconnect the different parts of the network and allow access to the fixed Public Switched Telephone Network (PSTN). The technology is hidden from view; it's incorporated in a number of tranceivers called Base Stations (BS). Every BS is located at a strategically selected place and covers a given area or cell - hence the name cellular communications. A number of adjacent cells grouped together form an area and the corresponding BSs communicate through a so called Mobile Switching Centre (MSC). The MSC is the heart of a cellular radio system. It is responsible for routing, or switching, calls from the originator to the destinator. It can be thought of managing the cell, being responsible for set-up, routing control and termination of the call, for management of inter-MSC hand over and supplementary services, and for collecting charging and accounting information. The MSC may be connected to other MSCs on the same network or to the PSTN.

The frequencies used vary according to the cellular network technology implemented. For GSM, 890-915 MHz range is used for transmission and 935-960 MHz for reception. The DCS techology uses frequencies in the 1800 MHz range while PCS in the 1900 MHz range.

Each cell has a number of channels associated with it. These are assigned to subscribers on demand. When a Mobile Station (MS) becomes 'active' it registers with the nearest BS. The corresponding MSC stores the information about that MS and its position. This information is used to direct incoming calls to the MS.

If during a call the MS moves to an adjacent cell then a change of frequency will necessarily occur - since adjacent cells never use the same channels. This procedure is called hand over and is the key to Mobile communications. As the MS is approaching the edge of a cell, the BS monitors the decrease in signal power. The strength of the signal is compared with adjacent cells and the call is handed over to the cell with the strongest signal.

During the switch, the line is lost for about 400 ms. When the MS is going from one area to another it registers itself to the new MSC. Its location information is updated, thus allowing MSs to be used outside their 'home' areas.

5.7.2. Data Communications

Data Communications is the exchange of data using existing communication networks. The term data covers a wide range of applications including File Transfer (FT), interconnection between Wide-Area-Networks (WAN), facsimile (fax), electronic mail, access to the internet and the World Wide Web (WWW).

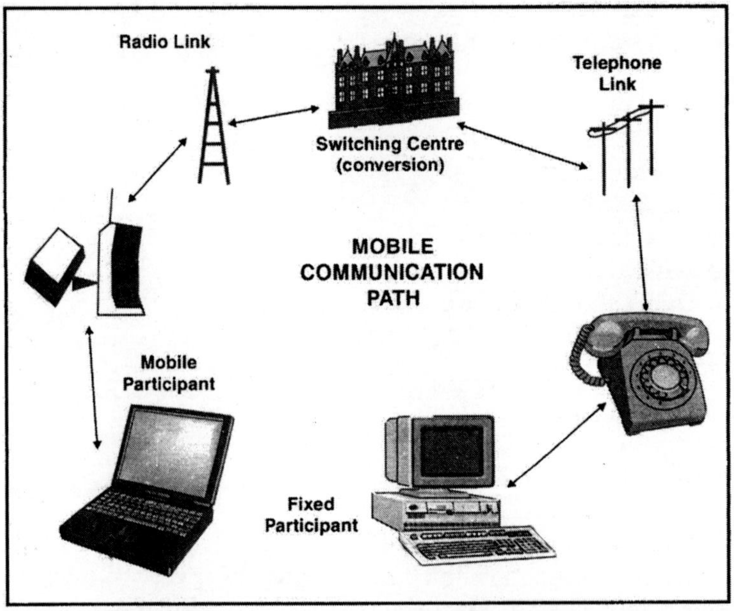

FIGURE 5.3. Mobile communications overview.

Data Communications have been achieved using a variety of networks such as PSTN, leased-lines and more recently ISDN (Integrated Services Data Network) and ATM (Asynchronous Transfer Mode)/Frame Relay. These networks are partly or totally analogue or digital using technologies such as circuit—switching, packet—switching etc.

1. Circuit switching implies that data from one user (sender) to another (receiver) has to follow a prespecified path. If a link to be used is busy, the message cannot be redirected, a property which causes many delays.

2. Packet switching is an attempt to make better utilisation of the existing network by splitting the message to be sent into packets. Each packet contains information about the sender, the receiver, the position of the packet in the message as well as part of the actual message. There are many protocols defining the way packets can be send from the sender to the receiver. The most widely used are the **Virtual Circuit-Switching** system, which implies that packets have to be sent through the same path, and the **Datagram** system which allows packets to be sent at various paths

depending on the network availability. Packet switching requires more equipment at the receiver, where reconstruction of the message will have to be done.

The introduction of mobility in data communications required a move from the Public Switched Data Network (PSDN) to other networks like the ones used by mobile phones. PCSI has come up with an idea called CDPD (Cellular Digital Packet Data) technology which uses the existing mobile network (frequencies used for mobile telephony).

Mobility implemented in data communications has a significant difference compared to voice communications. Mobile phones allow the user to move around and talk at the same time; the loss of the connection for 400ms during the hand over is undetectable by the user. When it comes to data, 400ms is not only detectable but causes huge distortion to the message. Therefore data can be transmitted from a mobile station under the assumption that it remains stable or within the same cell.

5.7.3. CDPD Technology (Cellular Digital Packet Data)

Today, the mobile data communications market is becoming dominated by a technology called CDPD.

There are other alternatives to this technology namely Circuit Switched Cellular, Specialised Mobile Radio and Wireless Data Networks. As can be seen from the table below the CDPD technology is much more advantageous than the others.

Parameters	Cellular Digital Packet Data (CDPD)	Circuit Switched Cellular	Specialized Mobile Radio (Extended)	Proprietary Wireless Data Networks
Speed	best	best	good	good
Security	best	better	good	better
Ubiquity	best	best	good	better
Cost of Service	best	better	better	good
Cost of Deployment	best	best	better	good
Mobility	best	good	better	good
Interoperability	best	good	good	better

CDPD's principle lies in the usage of the idle time in between existing voice signals that are being sent across the cellular networks. The major advantage of this system is the fact that the idle time is not chargeable and so the cost of data transmission is very low. This may be regarded as the most important consideration by business individuals.

CDPD networks allow fixed or mobile users to connect to the network across a fixed link and a packet switched system respectively. Fixed users have a fixed physical link to the CDPD network. In the case of a mobile end user, the user can, if CDPD network facilities are non-existent, connect to existing circuit switched networks and transmit data via these networks. This is known as Circuit Switched CDPD (CS-CDPD).

Service coverage is a fundamental element of providing effective wireless solutions to users and using this method achieves this objective. **Where CDPD is available data is split into packets and a packet switched network protocol is used to transport the packets across the network. This may be of either Datagram or Virtual Circuit Switching form.**

The data packets are inserted on momentarily unoccupied voice frequencies during the idle time on the voice signals. CDPD networks have a network hierarchy with each level of the hierarchy doing its own specified tasks.

The hierarchy consists of the following levels :

1. Mobile End User Interface

Using a single device such as a Personal Digital Assistant or personal computer which have been connected to a Radio Frequency (RF) Modem which is specially adapted with the antennae required to transmit data on the cellular network, the mobile end user can transmit both data and voice signals. Voice signals are transmitted via a mobile phone connected to the RF Modem Unit.

2. Mobile Data Base Station (MDBS)

In each cell of the cellular reception area, there is a Mobile Data Base Station (MDBS) which is responsible for detection of idle time in voice channels, for relaying data between the mobile units and the Mobile Data Intermediate Systems (MDIS), sending of packets of data onto the appropriate unoccupied frequencies as well as receiving data packets and passing them to the appropriate Mobile end user within its domain.

3. Mobile Data Intermediate Systems (MDIS)

Groups of MDBS that control each cell in the cellular network reception area are connected to a higher level entity in the network hierarchy, the Mobile Data Intermediate Systems. Connection is made via a wideband trunk cable. Data packets are then relayed by MDBS to and from mobile end users and MDIS.

4. Intermediate Systems (IS)

MDIS are interconnected to these IS which form the backbone of the CDPD system. These systems are unaware of mobility of end-users, as this is hidden by lower levels of the network hierarchy. The ISs are the systems that provide the CDPD interface to the various computer and phone networks.

The IS's relay data between MDIS's and other IS's throughout the network. They can be connected to routers that support Internet and Open Systems Interconnection Connectionless Network Services (OSI-CLNS), to allow access to other cellular carriers and external land based networks.

5.7.3.1. CDPD Network Reliability

There are some actions that are necessary in order to obtain reliability over a network.

1. **User Authentication** : The procedure which checks if the identity of the subscriber transferred over the radio path corresponds with the details held in the network.

2. **User Anonymity** : Instead of the actual directory telephone number, **the International Mobile Subscriber Identity (IMSI)** number is used within the network to uniquely identify a mobile subscriber.

3. **Fraud Prevention** : Protection against impersonation of authorised users and fraudulent use of the network is required.

4. **Protection of User Data** : All the signals within the network are encrypted and the identification key is never transmitted through the air. This ensures maximum network and data security.

CDPD follows the OSI standard model for packet switched data communications. The CDPD architecture extends across layers one, two and three of the OSI layer model. The mobile end users handle the layer 4 functions (transport) and higher layers of the OSI model such as user interface.

5.7.4. Applications of Mobile Computing

The question that always arises when a business is thinking of buying a mobile device is "Will it be worth to buy it ?"

In many fields of work, the ability to keep on the move is vital in order to utilise time efficiently. Efficient utilisation of resources (ie: staff) can mean substantial savings in transportation costs and other non quantifyable costs such as increased customer attention, impact of on site maintenance and improved intercommunication within the business.

The importance of Mobile Computers has been highlighted in many fields of which a few are described below :

1. **For Estate Agents :** Estate agents can work either at home or out in the field. With mobile computers they can be more productive. They can obtain current real estate information by accessing multiple listing services, which they can do from home, office or car when out with clients. They can provide clients with immediate feedback regarding specific homes or neighbourhoods, and with faster loan approvals, since applications can be submitted on the spot. Therefore, mobile computers allow them to devote more time to clients.

2. **Emergency Services :** Ability to receive information on the move is vital where the emergency services are involved. Information regarding the address, type and other details of an incident can be dispatched quickly, via a CDPD system using mobile computers, to one or several appropriate mobile units which are in the vicinity of the incident.

Here the reliability and security implemented in the CDPD system would be of great advantage.

3. **In courts :** Defense counsels can take mobile computers in court. When the opposing counsel references a case which they are not familiar, they can use the computer to get direct, real-time access to on-line legal database services, where they can gather information on the case and related precedents. Therefore mobile computers allow immediate access to a wealth of information, making people better informed and prepared.

4. **In companies :** Managers can use mobile computers in, say, critical presentations to major customers. They can access the latest market share information. They can revise the presentation to take advantage of this information. They can communicate with the office about possible new offers and call meetings for discussing responds to the new proposals. Therefore, mobile computers can leverage competitive advantages.

5. **Stock Information Control :** In environments where access to stock is very limited *i.e.* factory warehouses. The use of small portable electronic databases accessed via a mobile computer would be ideal.

Data collated could be directly written to a central database, via a CDPD network, which holds all stock information hence the need for transfer of data to the central computer at a later date is not necessary. This ensures that from the time that a stock count is completed, there is no inconsistency between the data input on the portable computers and the central database.

6. **Credit Card Verification :** At Point of Sale (POS) terminals in shops and supermarkets, when customers use credit cards for transactions, the intercommunication required between the bank central computer and the POS terminal, in order to effect verification of the card usage, can take place quickly and securely over cellular channels using a mobile computer unit. This can speed up the transaction process and relieve congestion at the POS terminals.

7. **Taxi/Truck Dispatch :** Using the idea of a centrally controlled dispatcher with several mobile units (taxis), mobile computing allows the taxis to be given full details of the dispatched job as well as allowing the taxis to communicate information about their whereabouts back to the central dispatch office. This system is also extremely useful in secure deliveries. This allows a

central computer to be able to track and recieve status information from all of its mobile secure delivery vans. Again, the security and reliabilty properties of the CDPD system shine through.

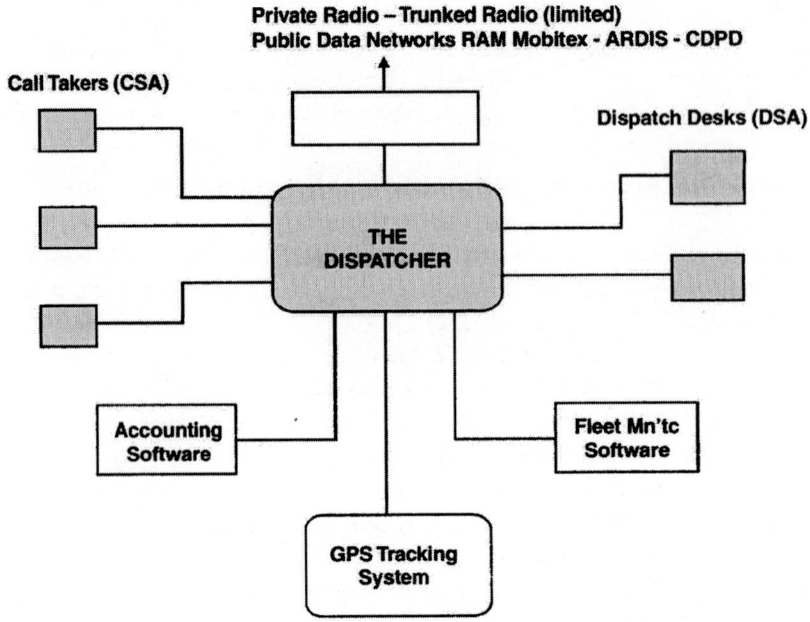

FIGURE 5.4. Taxi dispatch network.

8. **Electronic Mail/Paging** : Usage of a mobile unit to send and read emails is a very useful asset for any business individual, as it allows him/her to keep in touch with any colleagues as well as any urgent developments that may affect their work. Access to the Internet, using mobile computing technology, allows the individual to have vast arrays of knowledge at his/her fingertips.

Paging is also achievable here, giving even more intercommunication capability between individuals, using a single mobile computer device.

5.7.5. The Future

With the rapid technological advancements in Artificial Intelligence, Integrated Circuitry and increases in Computer Processor speeds, the future of mobile computing looks increasingly exciting.

With the emphasis increasingly on compact, small mobile computers, it may also be possible to have all the practicality of a mobile computer in the size of a hand held organizer or even smaller.

5.8. INTRODUCTION TO WAP (WIRELESS APPLICATION PROTOCOL)

With the growing need for wireless Internet access, the web designers are now encountering a new challenge in creating support for mobility. Also, mobile access is based on different demand of information than conventional web usage. Getting bus time tables right in the mobile phone while waiting for a next bus on the way to work, reading e-mails by GSM before entering the office and checking the web pages of a customer firm just before first meeting are all examples of the need of mobile access to Internet and other web services.

Wireless Application Protocol enables easy fast delivery of relevant information and services to mobile users with wireless terminals with limited displays and data transfer capabilities. **It is a specification for a set of communication protocols to standardize the way in which cellular devices use Internet access, including www, newsgroups, e-mail and IRC.**

5.8.1. WAP Forum

In February 1998, the founding members of WAP Forum published the first draft of the WAP specification. The goal of WAP Forum is to bring together companies for creating a global wireless protocol specification which will operate both with differing types of wireless device and network technologies. It was established in 1997 by Ericsson, Motorola, Nokia and Unwired Planet and now has over than 90 members, which together represent over 90 % of the global handset market.

The objectives of the WAP Forum are :

(*i*) To bring Internet content and advanced data services to digital cellurar phones and other wireless terminals.

(*ii*) To create a global wireless protocol specification that will work across differing wireless network technologies.

(*iii*) To enable the creation of content and applications that scale across a very wide range of bearer networks and device types.

(*iv*) To embrace and extend existing standards and technology wherever appropriate.

After publishing the first draft, WAP Forum held many open meetings to discuss the draft specification. The completed WAP version 1.0 specification is based on proposals suggested in meetings held in Miami and Tokyo, as well as feedback got via the WAP web site. Now finalished version 1.0 specification is available via WAP Forum web site with proposals for version 1.1.

5.8.2. Why do We Need WAP ?

Having the performance and data transfer capacities of the common desktop computers in mind, the web designers constructed the Internet technology for devices as powerful as those computers. Hand-held wireless devices present more a constrained computing environment with less powerful CPU's, less memory, restricted power consumption, smaller displays and different input devices. Similarly, wireless data networks have less bandwidth, more latency, less connection stability and less predictable availability than conventional wired networks. Also, the operators aim to provide the users advanced services in more a usable and attractive way.

The WAP specification meets the mobile network characteristics and operators needs by filling the special requirements of hand held wireless terminals by using existing network technology and introducing new technology where appropriate.

5.8.3. WAP Architecture Overview

The WAP programming model is similar to the Internet model. This provides some benefits, including the familiar architecture and programming environment as well as the ability to use existing tools. Not only Internet standards have been used where possible, but also optimisations and extensions have been made.

WAP specifies two essential elements of wireless communication: an end-to-end application protocol and an application environment based on a browser. The application protocol is a layered communication protocol that is embedded in each WAP user agent. The network side includes a server component implementing the other end of the protocol that is capable of communicating with any WAP user agents. The role of the server component is also to act as a gateway to route the request of a user agent to an application server. Physically the gateway can be located in a telecom or a computer network, in order to build a bridge between the two different networks.

WAP provides a standard to enable running the same application in different user agents. The standard consists of a browser and a script interpreter. The browser acts like a conventional web browser but instead of HTML, it understands content written in Wireless Markup Language or briefly, WML. Also instead of JavaScript interpreter, WAP browser has a build in script interpreter for running applications written in a script language called WMLScript. Furthermore the browser includes some libraries to allow the application to use certain services of the user agents. Both WML and WMLScript are designed to be used in wireless networks with narrowband data transfer rates, so they are both binary encoded to optimise the data transmission efficiency.

5.8.4. WAP Protocol Stack

WAP specifies an architecture based on layers that follows the OSI model fairly closely. The WAP model, or stack as it is commonly known, is illustrated in Figure 5.5.

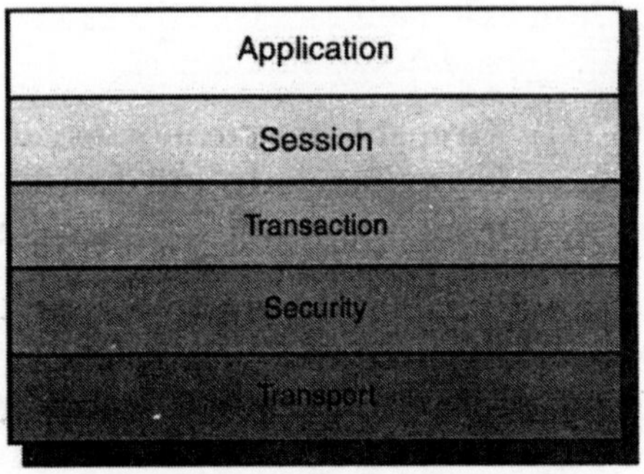

FIGURE 5.5. The WAP model

Application Layer

WAP's application layer is the Wireless Application Environment (WAE). WAE directly supports WAP application development with Wireless Markup Language (WML) instead of HTML and WMLScript instead of JavaScript. WAE also includes the Wireless Telephony Application Interface (WTAI, or WTA for short) that provides a programming interface to telephones for initiating calls, sending text messages, and other networking capability.

Session Layer

WAP's session layer is the Wireless Session Protocol (WSP). WSP is the equivalent to HTTP for WAP browsers. WAP involves browsers and servers just like the Web, but HTTP was not a practical choice for WAP because of its relative inefficiency on the wire. WSP conserves precious bandwidth on wireless links; in particular, WSP works with relatively compact binary data where HTTP works mainly with text data.

Transaction, Security, and Transport Layers

These three protocols can be thought of as "glue layers" in WAP:

- Wireless Transaction Protocol (WTP)
- Wireless Transaction Layer Security (WTLS)
- Wireless Datagram Protocol (WDP)

WTP provides transaction-level services for both reliable and unreliable transports. It prevents duplicate copies of packets from being received by a destination, and it supports retransmission, if necessary, in cases where packets are dropped. In this respect, WTP is analogous to TCP. However, WTP also differs from TCP. WTP is essentially a pared-down TCP that squeezes some extra performance from the network.

WTLS provides authentication and encryption functionality analogous to Secure Sockets Layer (SSL) in Web networking. Like SSL, WTLS is optional and used only when the content server requires it.

WDP implements an abstraction layer to lower-level network protocols; it performs functions similar to UDP. WDP is the bottom layer of the WAP stack, but it does not implement physical or data link capability. To build a complete network service, the WAP stack must be implemented on some low-level *legacy* interface not technically part of the model. These interfaces, called *bearer services* or *bearers*, can be IP-based or non-IP based.

Bearer Interfaces

WAP supports dial-up networking using IP and Point-to-Point Protocol (PPP) as the bearer interface. It also supports Short Message Service (SMS) and General Packet Radio System (GPRS). SMS passes text and binary data between digital phones. GPRS is a relatively new technology that implements faster, "always-on" connections for wireless devices; GPRS actually runs on top of IP.

5.8.5. Mobile Information Devices

A special kind of devices needed to perform the M-commerce. Let us discuss some of these devices :

1. Portable Computers
2. Hybrid Pen Computers
3. Personal Digital Assistance (PDA)
4. Digital Assistance
5. Personal Communicators
6. Palmtops

By using these devices we can perform the M-commerce. By using the wire less network we can connect our devices (enabled with mobile technology) with Internet to perform the M-commerce. Let us see some detail of these devices as follows :

1. Portable Computers : Basically Laptops, note books and handhelds comes under portable computers. Laptops are needed to run some heavy applications. Notebook has the capacity to run word data and manipulate spread sheets. Handhelds, like mobile phones has limited processing powers. But nowadays mobile phones also becomes very advance and can work like small computers.

2. Hybrid Pen Computers : A pen-based interface is clearly more appropriate in many situations. Pen computers are often equipped with wireless communications. Basically in hybrid pen computers a pen is provided which works as mouse.

3. PDAs (Personal Digital Assistance) : PDA means a computer which can be adjusted in pocket also. Basically it is a smallest possible computer and still companies are working to reduce the size.

4. Digital Assistants : Digital assistant combine personal information management with wireless voice, data and fax communication. Digital assistants provides a wide array of information-management tools, note pad and sketch pad functionality.

5. Personal Communicators : Basically it is a PDA which is a combination of cellular telephone with a pen computer's user interface. Simon, designed by IBM is the example of it. Personal communicators offers many services like fax transmission and reception, personal information management, email without attachments etc.

6. Palmtops computers provides higher functionality and more closely resemble what is available in desktop computers. In palmtop software is built into the ROM (ready only memory). Palmtops have combined functionality of a personal organizer and communication terminal to provide access to e-mail.

SUMMARY

In this chapter, we have seen the subset of e-commerce, that is M-commerce. In M-commerce business-to-consumer transactions conducted from a mobile device. Basically M-commerce is the e-commerce over mobile. We have seen the application of M-commerce in different fields. In this chapter a basic overview of cellular network architecture was discussed. Portable computers, hybrid pen computers, PDA's, Digital Assistance, Personal Communicators and Palmtops are mobile devices, through which we can perform M-commerce.

REVIEW QUESTIONS

1. What is M-commerce and how it is different from e-commerce.
2. How M-commerce can help us to uplift our living standards.
3. Discuss some driving force for M-commerce, what are various limiting factors in M-commerce ?
4. How mobile computing is helpful ?
5. Now a days, almost everybody is using mobile-phones. What kind of basic cellular network architecture exists to send and receiver the phone calls.
6. "Mobile data communication market is becoming dominated by a technology called CDPD". Justify this statement by comparing CDPD with others.
7. How WAP architecture is different from TCP/IP.
8. What is WAP and what is the need of it ?
9. Discuss the Layered structure of WAP.
10. "M-Commerce is possible through mobile devices only." Discuss the various mobile devices.

❑❑❑

6

CHAPTER

WEB SECURITY

INSIDE THIS CHAPTER

6.1. E-Commerce Security Issue
6.2. Web Security
6.3. Client Server Technology
6.4. Firewalls and Network Security

6.1. E-COMMERCE SECURITY ISSUE

While many security issues in e-commerce are the same as general security issues, some of them are specific for the kind of software used by e-commerce businesses: databases, in particular databases which are accessed remotely, online forms and shopping carts. Below we consider these specific vulnerabilities.

6.1.1. Database Vulnerabilities

More common vulnerabilities in e-commerce are caused remotely accessed databases. Below are some examples :

(*i*) Running SQL queries based on data entered by the user may allow a malicious user to append their own query to the one that is supposed to be executed. Similar things can be done, in addition to online forms, with URL rewriting and cookies. Note that the user can easily type in any URL into the browser window, including a URL which contains an extra query. The user can also easily alter a cookie which resides on the user's computer.

To avoid these vulnerabilities, one has to check the data entered by the user, as well as URLs and cookies, to make sure that they are of the correct format. One can also restrict privileges of the process running search queries so that it is not allowed to remove data from the database. **One also has to be careful not to expose table names in the text of online forms, since this attack requires the hacker to know the table name in order to delete the information from it.**

(*ii*) Another group of database security issues comes from exposing database servers to hackers. **To prevent unauthorized access, database servers should be inside a firewall.** It is also important that database passwords and user names are not exposed in web pages that a user can see (not only in plain text, but also in hidden fields, and in Java script code). Such passwords and user names should be transmitted encrypted.

(*iii*) Since many employees of an online business have access to the database, their account names and passwords should be kept secure. Also, all code that contains passwords for database access (such as Java code of applets and servlets) should be stored securely.

6.1.2. Shopping Cart Vulnerabilities

Numerous vulnerabilities have been discovered in shopping carts, both commercially produced and "home made" ones:

(*i*) **Price manipulation :** If a price of the product is passed as a hidden parameter in an online form and the value of the parameter is used to determine the charge on the credit card, then a customer can easily replace the HTML page with the form by their own page, where all the information is the same, but the price is changed to a lower one.

(*ii*) **Transaction modification** uses a similar idea. If the merchant's site redirects the user to another website for credit card verification or sends a request for the charge to the credit card company, this is often done via a POST request. The request may contain the amount of the credit card charge or the kind of a transaction (some transactions may be specified as "test", in which case no actual charge takes place). The user may modify the amount of charge or mark their transaction as "test".

(*iii*) **Exposing Merchant's ID :** When the merchant sends a credit card charge request, the request may contain the merchant's ID. If the ID is exposed in the website forms, a hacker may use this ID to send their own requests to the credit card company. The request may include canceling a transaction, thus crediting money to the credit card. Hackers also may use merchant's ID to check if a credit card number is valid.

(*iv*) **Other things that may be exposed by poorly written shopping cart code are :** IP addresses of database servers, user name and password for database access, names of database tables and columns. All this may cause theft, damage, or modification of the data stored in the database.

(*v*) **Another group of break-ins is related to so-called back door access to a shopping cart :** Shopping carts come with passwords that allow the owner to change settings or access information in the shopping cart or the database. Initially these passwords are set to default. Once the default becomes known to hackers, the software can be easily accessed and changed to serve the purposes of the hacking.

6.1.3. Customer's Passwords

Many online businesses use customer's passwords to authorize access to sensitive data (such as the order information). However, one has to be extremely careful with how the password system works. Here are potential problems :

(*i*) **People Tend to Forget Their Passwords.** A forgotten password needs to be reset. However, resetting a password should be done securely. The confirmation should be sent by e-mail, and the new password should not be used until the customer replies to the e-mail. One of the recent cases of security breaches involving changing passwords was discovered at eBay.

(*ii*) **Another possibility for stealing passwords is to "impersonate" a merchant and to request the user's password.** In general, a business website may be better off without customer's passwords, and should not rely on them for protection of sensitive information.

6.2. WEB SECURITY

Unfortunately, there's a lot to worry about. There are security risks that affect Web servers, the local area networks that host Web sites, and even innocent users of Web browsers.

The risks are most severe from the Webmaster's perspective. The moment we install a Web server at our site, we have opened a window into our local network that the entire Internet can peer through. Most visitors are content to window shop, but a few will try to to peep at things we don't intend for public consumption. The results can range from the merely embarrassing, for instance the discovery one morning that our site's home page has been replaced by an obscene parody, to the damaging, for example the theft of your entire database of customer information.

The open architecture of Web servers allows arbitrary CGI scripts to be executed on the server's side of the connection in response to remote requests. Any CGI script installed at your site may contain bugs, and every such bug is a potential security hole.

From the point of view of the network administrator, a Web server represents yet another potential hole in your local network's security. The general goal of network security is to keep strangers out. Yet the point of a Web site is to provide the world with controlled access to your network. Drawing the line can be difficult. A poorly configured Web server can punch a hole in the most carefully designed firewall system. A poorly configured firewall can make a Web site impossible to use. Things get particularly complicated in an intranet environment, where the Web server must typically be configured to recognize and authenticate various groups of users, each with distinct access privileges.

Finally, both end-users and Web administrators need to worry about the confidentiality of the data transmitted across the Web. The TCP/IP protocol was not designed with security in mind; hence it is vulnerable to network hacking. When confidential documents are transmitted from the Web server to the browser, or when the end-user sends private information back to the server inside a fill-out form, someone may be listening in.

6.2.1. Types of Web Security Risks

There are basically three overlapping types of risk :

1. **Bugs or misconfiguration problems in the Web server** that allow unauthorized remote users to :
   (*a*) Steal confidential documents not intended for their eyes.
   (*b*) Execute commands on the server host machine, allowing them to modify the system.
   (*c*) Gain information about the Web server's host machine that will allow them to break into the system.
   (*d*) Launch denial-of-service attacks, rendering the machine temporarily unusable.
2. **Browser-side risks, including :**
   - Active content that crashes the browser, damages the user's system, breaches the user's privacy, or merely creates an annoyance.
   - The misuse of personal information knowingly or unkowingly provided by the end-user.
3. **Interception of network data sent from browser to server or vice versa via network hackers.** Hackers can operate from any point on the pathway between browser and server including :
   (*a*) The network on the browser's side of the connection.
   (*b*) The network on the server's side of the connection (including intranets).

(*c*) The end-user's Internet service provider (ISP).

(*d*) The server's ISP.

(*e*) Either ISPs' regional access provider.

It's important to realize that "secure" browsers and servers are only designed to protect confidential information against network hackers. Without system security on both browser and server sides, confidential documents are vulnerable to interception.

6.2.2. Web Security Precautions

If we are a Webmaster, system administrator, or are otherwise involved with the administration of a network, the single most important step we can take to increase our site's security is to create a written security policy. This security policy should succinctly lay out our organization's policies with regard to :

(*a*) who is allowed to use the system

(*b*) when they are allowed to use it

(*c*) what they are allowed to do (different groups may be granted different levels of access)

(*d*) procedures for granting access to the system

(*e*) procedures for revoking access (e.g. when an employee leaves)

(*f*) what constitutes acceptable use of the system

(*g*) remote and local login methods

(*h*) system monitoring procedures

(*i*) protocols for responding to suspected security breaches

This policy need not be anything fancy. It need only be a succinct summary of how the information system work, reflecting your organization's technological and political realities. There are several benefits to having a written security policy:

1. We ourself will understand what is not permitted on the system. If we don't have a clear picture of what is permitted, we can never be sure when a violation has occurred.
2. Others in our organization will understand what the security policy is. The written policy raises the level of security consciousness, and provides a focal point for discussion.
3. The security policy serves as a requirements document against which technical solutions can be judged. This helps guard against the "buy first, ask questions later" syndrome.

6.3. CLIENT SERVER TECHNOLOGY

Client/Server technology is a means for separating the functions of an application into two or more distinct parts. The client presents and manipulates data on the desktop computer. The server acts like a mainframe to store and retrieve protected data. Together each machine can perform the duties it is best at.

Among the characteristics of a client/server architecture are the following :

(*a*) The client and server can be distinguished from one another by the differences in tasks they perform.

(*b*) The client and server usually operate on different computer platforms.

(*c*) Either the client or server may be upgraded without affecting the other. Clients may connect to one or more servers; servers may connect to multiple clients concurrently.

(*d*) Clients always initiate the dialogue by requesting a service.

6.3.1. Overview of the Client Server Technology

Here's a little map of Client server technologies. The client computer (the hardware sitting perhaps on our desktop) communicates across the Internet (by modem or Ethernet) with a server computer (usually at a remote site).

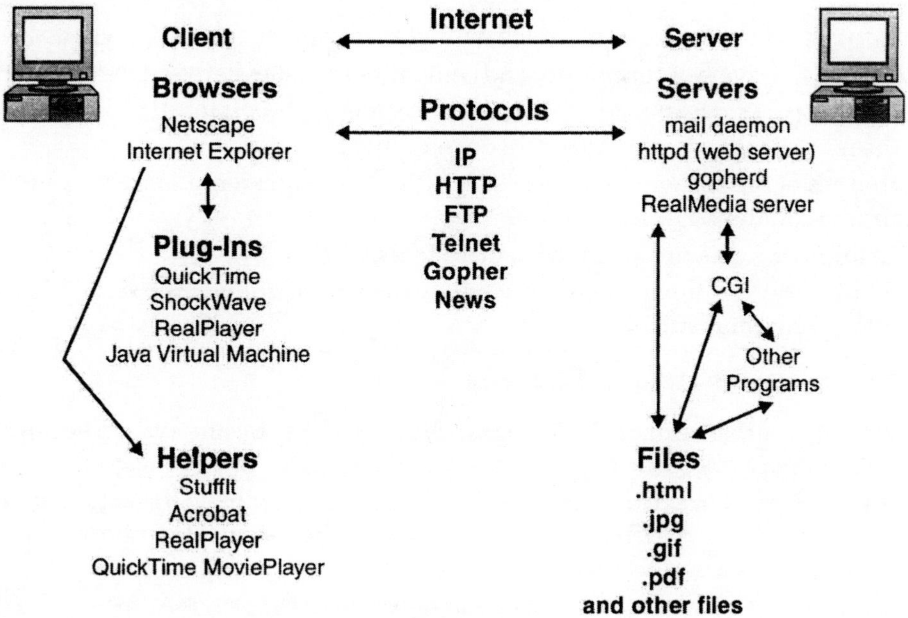

FIGURE 6.1.

The client software on the client computer is a web browser or a telnet program. It contacts software called servers on the server computer. These servers are usually daemons (a name for software that is always running). The client software and the server software communicate by means of a protocol (an established pattern of behaviours and responses).

So for example, a web browser may contact a web server using the HTTP protocol. The browser sends a request for a particular web page. The web server logs the request and then finds the correct file and sends it back across the Internet to the browser. The browser receives the file and interprets it for the user by supplying images, format and so on.

Helper Applications run on the client's side to help show files that require special handling, such as movies, or compressed files, or Acrobat (.pdf) files. Plug-Ins are software that actually run inside the web brower to display special files.

On the server side, Common Gateway Interface (CGI) programs can access various files and other programs on the server computer and send dynamically created web pages back across the Internet. For example, this is the way search engines work: each search creates a new page that doesn't already exist as a file.

6.3.2. Benefits

The following are what are considered to be the benefits of client/server computing.

(*a*) Many times easier to implement client/server than change a legacy application.

(*b*) New technology and the move to rapid application development such as object oriented technology

(c) Long term cost benefits for development and support. Easy to add new hardware to support new systems such as document imaging and video teleconferencing which would not be feasible or cost efficient in a mainframe environment. Can implement multiple vendor software tools for each application.

6.3.3. Concerns

The following are concerns for companies willing to implement client/server computing.

(i) There are many layers of complexity and compatibility issues between the client and server.

(ii) Cost will rise in the short term since we will need to maintain the old system (mainframe) and the new client server architecture development.

(iii) Capabilities of the software such as security and management tools are not as mature as mainframe counterparts.

(iv) Takes time to become proficient with these tools.

(v) Information System departments may have problems at giving up control of a centralized computing environment.

6.3.4. Client server Network Security

Now a days client server security is the bigest challenge faces by any system administrator. In general client server network security problems can be categories in three ways :

(a) **Physical Security Threats :** In this case unauthorized user have the access of our system. Once they have physical access then they can do, whatever they want. Unauthorized accesser can have access of both hardware and software.

(b) **Software Security Threats :** In this case a poorly written program can do the damage of highest level. Through these codes intruder may have highest level of access possible and cold delete the entire file system, or create a new account or password file resulting in incalculable damage. For example "rlogin" hold in the IBM RS-6000 workstations, which enabled a cracker to create a "root" shell or super user access mode.

(c) **Inconsistent Usage Threats :** Some times due to some specific combination of software and hardware, system becomes very open to security point of view.

We can use various protection method at various levels to overcome these security threats such as :

(1) **Trsuted Base Security :** In which we trust everyone can do nothing extra for protection. But now a days this case is obselete.

(2) **Applying Password Schemes :** We can apply passwords at various level in clients and server computers to restrict unauthorized user to intract with servers as well as client.

(3) **Applying Biometric passwords** may be the key to stop the unauthorize use of computers.

(4) **Applying voice matching passwords** is the key now a days to authenticate the users.

Even we can restrict physical access through various approaches like biometric and voice matching etc.

6.3.5. Emerging Client Server Security Threats

Now, we've covered enough background information on client server that we can actually get into the security aspects of all of this. First of all, we'll get into the types of threats there are against networked computers, and then some things that can be done to protect ourself against various threats. Some main threats as follows :

1. Denial-of-Service

DoS *(Denial-of-Service) attacks are probably the common, and most difficult to address. These are the common, because they're very easy to launch, difficult (sometimes impossible) to track, and it isn't easy to refuse the requests of the attacker, without also refusing legitimate requests for service.*

Some things that can be done to reduce the risk of being stung by a denial of service attack include :

(a) Not running your visible-to-the-world servers at a level too close to capacity.

(b) Using packet filtering to prevent obviously forged packets from entering into your network address space.

(c) Keeping up-to-date on security-related patches for your hosts' operating systems.

2. Unauthorized Access

"Unauthorized access" is a very high-level term that can refer to a number of different sorts of attacks. The goal of these attacks is to access some resource that our machine should not provide the attacker. For example, a host might be a web server, and should provide anyone with requested web pages. However, that host should not provide command shell access without being sure that the person making such a request is someone who should get it, such as a local administrator.

3. Executing Commands Illegally

It's obviously undesirable for an unknown and untrusted person to be able to execute commands on your server machines. There are two main classifications of the severity of this problem: normal user access, and administrator access. A normal user can do a number of things on a system (such as read files, mail them to other people, etc.) that an attacker should not be able to do. This might, then, be all the access that an attacker needs. On the other hand, an attacker might wish to make configuration changes to a host (perhaps changing its IP address, putting a start-up script in place to cause the machine to shut down every time it's started, or something similar). In this case, the attacker will need to gain administrator privileges on the host.

4. Confidentiality Breaches

We need to examine the threat model: what is it that we are trying to protect ourself against ? There is certain information that could be quite damaging if it fell into the hands of a competitor, an enemy, or the public. In these cases, it's possible that compromise of a normal user's account on the machine can be enough to cause damage (perhaps in the form of public relation or obtaining information that can be used against the company, etc.).

5. Destructive Behavior

Among the destructive sorts of break-ins and attacks, there are two major categories.

(*a*) Data Diddling : The data diddler is likely the worst sort, since the fact of a break-in might not be immediately obvious. Perhaps he's toying with the numbers in your spreadsheets, or changing the dates in your projections and plans. Maybe he's changing the account numbers for the auto-deposit of certain paychecks. In any case, rare is the case when you'll come in to work one day, and simply know that something is wrong. An accounting procedure might turn up a discrepancy in the books three or four months after the fact. Trying to track the problem down will certainly be difficult, and once *that* problem is discovered, how can any of your numbers from that time period be trusted? How far back do you have to go before you think that your data is safe ?

(*b*) Data Destruction : Some of those perpetrate attacks are simply twisted jerks who like to delete things. In these cases, the impact on our computing capability—and consequently our business—can be nothing less than if a fire or other disaster caused our computing equipment to be completely destroyed.

6.4. FIREWALLS AND NETWORK SECURITY

"Firewall is a software which enforces an access control policy between an internal network in an organization and the rest of the Internet. It may also be used for access control between two parts of the same internal network. A firewall blocks some Internet traffic and permits some other traffic".

Some firewalls are more restrictive than others. It's important to understand that a firewall implements a policy, so it's only as good as the policy it implements. If the policy is inconsistent or not well thought of, then a firewall may not be a good protection.

The main thing firewalls protect against is unauthorized login. In addition they may block some or all of the outside traffic, while permitting all or most of the traffic from the inside to the outside. The traffic that usually is permitted is e-mail (both to and from the network) and HTTP access from the inside to the outside of the firewall. Protocols that may or may not permitted are FTP, SSL, database connections, and so on. In addition to access control, firewalls keep statistics of the Internet packages and requests, and may be configured to issue a warning if some suspicious activity is going on.

There are two basic types of firewalls: *network layer* and *application layer* firewalls. Recently the distinction between the two is becoming somewhat blurry.

6.4.1. Importance of Firewall

The Internet, like any other society, is plagued with the kind of jerks who enjoy the electronic equivalent of writing on other people's walls with spray paint, tearing their mailboxes off, or just sitting in the street blowing their car horns. Some people try to get real work done over the Internet, and others have sensitive or proprietary data they must protect. Usually, **a firewall's purpose is to keep the jerks out of our network while still letting us get our job done.**

Many traditional-style corporations and data centers have computing security policies and practices that must be followed. In a case where a company's policies dictate how data must be protected, a firewall is very important, since it is the embodiment of the corporate policy. Frequently, the hardest part of hooking to the Internet, if we are a large company, is not justifying the expense or effort, but convincing management that it's safe to do so. **A firewall provides not only real security—it often plays an important role as a security blanket for management.**

Lastly, a firewall can act as your corporate "ambassador" to the Internet. Many corporations use their firewall systems as a place to store public information about corporate products and services, files to download, bug-fixes, and so forth. Several of these systems have become important parts of the Internet service structure and have reflected well on their organizational sponsors. Note that while this is historically true, most organizations now place public information on a Web server, often protected by a firewall, but not normally on the firewall itself.

6.4.2. Firewall Protection

Some firewalls permit only email traffic through them, thereby protecting the network against any attacks other than attacks against the email service. Other firewalls provide less strict protections, and block services that are known to be problems.

Generally, firewalls are configured to protect against unauthenticated interactive logins from the "outside" world. This, more than anything, helps prevent intruders from logging into machines on your network. More elaborate firewalls block traffic from the outside to the inside, but permit users on the inside to communicate freely with the outside. The firewall can protect you against any type of network-borne attack if you unplug it.

Firewalls are also important since they can provide a single "choke point" where security and audit can be imposed. Unlike in a situation where a computer system is being attacked by someone dialing in with a modem, the firewall can act as an effective "phone tap" and tracing tool. Firewalls provide an important logging and auditing function; often they provide summaries to the administrator about what kinds and amount of traffic passed through it, how many attempts there were to break into it, etc.

Because of this, firewall logs are critically important data. They can be used as evidence in a court of law in most countries. We should safeguard, analyze and protect our firewall logs accordingly.

This is an important point: providing this "choke point" can serve the same purpose on our network as a guarded gate can for your site's physical premises. That means anytime we have a change in "zones" or levels of sensitivity, such a checkpoint is appropriate. A company rarely has only an outside gate and no receptionist or security staff to check badges on the way in. If there are layers of security on our site, it's reasonable to expect layers of security on your network.

6.4.3. Limitations of Firewall

Firewalls can't protect against attacks that don't go through the firewall. Many corporations that connect to the Internet are very concerned about proprietary data leaking out of the company through that route. Unfortunately for those concerned, a magnetic tape, compact disc, DVD, or USB flash drives can just as effectively be used to export data. Many organizations that are terrified (at a management level) of Internet connections have no coherent policy about how dial-in access via modems should be protected. It's silly to build a six-foot thick steel door when you live in a wooden house, but there are a lot of organizations out there buying expensive firewalls and neglecting the numerous other back-doors into their network. *For a firewall to work, it must be a part of a consistent overall organizational security architecture.* Firewall policies must be realistic and reflect the level of security in the entire network. For example, a site with top secret or classified data doesn't need a firewall at all: they shouldn't be hooking up to the Internet in the first place, or the systems with the really secret data should be isolated from the rest of the corporate network.

Another thing a firewall can't really protect you against is traitors or idiots inside your network. While an industrial spy might export information through our firewall, he's just as likely to export it through a telephone, FAX machine, or Compact Disc. CDs are a far more likely means for information to leak from our organization than a firewall. Firewalls also cannot protect us against stupidity. Users who reveal sensitive information over the telephone are good targets for social engineering; an attacker may be able to break into our network by completely bypassing our firewall, if he can find a "helpful" employee inside who can be fooled into giving access to a modem pool. Before deciding this isn't a problem in our organization, ask ourself how much trouble a contractor has getting logged into the network or how much difficulty a user who forgot his password has getting it reset. If the people on the help desk believe that every call is internal, we have a problem that can't be fixed by tightening controls on the firewalls.

Lastly, firewalls can't protect against bad things being allowed through them. For instance, many Trojan Horses use the Internet Relay Chat (IRC) protocol to allow an attacker to control a compromised internal host from a public IRC server. **If we allow any internal system to connect to any external system, then our firewall will provide no protection from this vector of attack.**

6.4.4. Types of Firewalls

There are three basic types of firewalls, and we'll consider each of them.

6.4.4.1. Application Gateways or Proxy Gateways

The first firewalls were application gateways, and are sometimes known as proxy gateways. These are made up of hosts that run special software to act as a proxy server. This software runs at the *Application Layer* of our old friend the ISO/OSI Reference Model, hence the name. Clients behind the firewall must be *proxitized* (that is, must know how to use the proxy, and be configured to do so) in order to use Internet services. Traditionally, these have been the most secure, because they don't allow anything to pass by default, but need to have the programs written and turned on in order to begin passing traffic.

These are also typically the slowest, because more processes need to be started in order to have a request serviced. Figure 6.2 shows a application gateway.

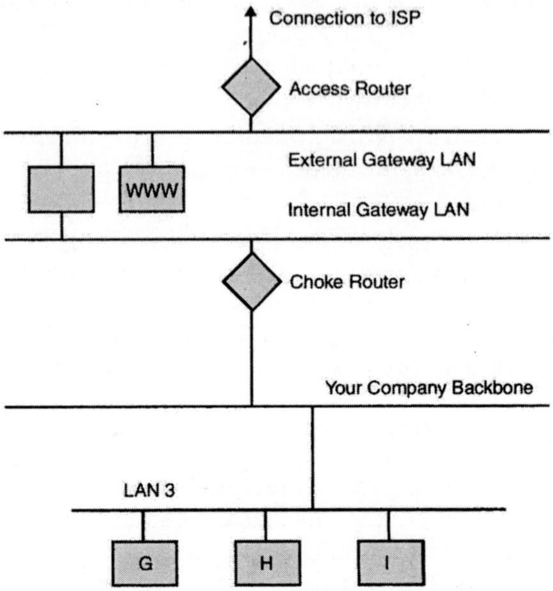

FIGURE 6.2. A sample application gateway.

6.4.4.2. Packet Filtering

Packet filtering is a technique whereby routers have *ACLs* (Access Control Lists) turned on. By default, a router will pass all traffic sent in and will do so without any sort of restrictions. Employing ACLs is a method for enforcing our security policy with regard to what sorts of access we allow the outside world to have to our internal network, and vice versa.

There is less overhead in packet filtering than with an application gateway, because the feature of access control is performed at a lower ISO/OSI layer (typically, the transport or session layer). Due to the lower overhead and the fact that packet filtering is done with routers, which are specialized computers optimized for tasks related to networking, a packet filtering gateway is often much faster than its application layer cousins. Figure 6.3 shows a packet filtering gateway.

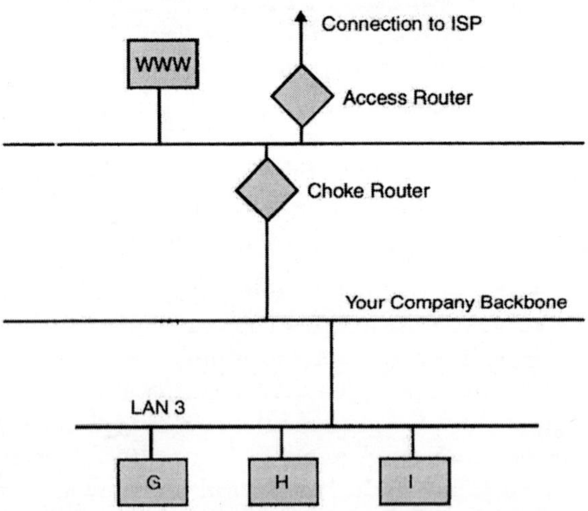

FIGURE 6.3. A sample packet filtering gateway.

Because we're working at a lower level, supporting new applications either comes automatically, or is a simple matter of allowing a specific packet type to pass through the gateway.

There are problems with this method, though. Remember, TCP/IP has absolutely no means of guaranteeing that the source address is really what it claims to be. As a result, we have to use layers of packet filters in order to localize the traffic. We can't get all the way down to the actual host, but with two layers of packet filters, we can differentiate between a packet that came from the Internet and one that came from our internal network. **We can identify which network the packet came from with certainty, but we can't get more specific than that.**

6.4.4.3. Hybrid Systems

In an attempt to have the security of the application layer gateways with the flexibility and speed of packet filtering, some vendors have created systems that use the principles of both.

In some of these systems, new connections must be authenticated and approved at the application layer. Once this has been done, the remainder of the connection is passed down to the session layer, where packet filters watch the connection to ensure that only packets that are part of an ongoing (already authenticated and approved) conversation are being passed.

Other possibilities include using both packet filtering and application layer proxies. The benefits here include providing a measure of protection against your machines that provide services to the Internet (such as a public web server), as well as provide the security of an application layer gateway to the internal network.

SUMMARY

Security is major concern in e-commerce. Every e-commerce activity is performed through web, so it is very important to insure the web security. Client server technology is bases of internet and it also have its own security concern which we have to overcome to insure secure transactions. Firewall is a combination of software and hardware, which ensures security of network at various level.

REVIEW QUESTIONS

1. "Security is major concern in e-commerce implementation." So discuss the various security issues in e-commerce.
2. How we can ensure a safe and secure e-commerce implementation through internet ?
3. "Working on internet is always risky but without it we cannot do most of our work". So what web security precautions we can have ?
4. What is the role of client server technology in internet ?
5. What is role of a firewall in network security ?
6. In what situation, which firewall should be used ?

❑❑❑

7

CHAPTER

ENCRYPTION

7.1. WORLD WIDE WEB SECURITY

INSIDE THIS CHAPTER

7.1. World Wide Web Security
7.2. Internet Transactions and Security
7.3. Encryption
7.4. Digital Certificate
7.5. Secure Socket Layer (SSL)
7.6. Authentication
7.7. Introduction to VPN

We know that almost all e-commerce and m-commerce activities can be performed through WWW services. Most of the transactions are done for important information and finance related. So security is the major thrust are for transaction take place on web.In order to counter the various types of security threats/attacks to transmitted electronic data, security services to ensure reliable and true worthy transmission of digital business data and messages are needed. The primary security services are divided into five categories :

- Confidentiality
- Integrity
- Authentication
- Non-repudiation
- Authorization
- Availability

Different organizations and individual may face different types of threats to the security of their information on the Internet. Some of the common risks/threats are :

1. Data may be stolen, corrupted misused, altered.
2. Attacks on hardware may occur that may render systems unable to operate properly.
3. Programs may be altered to cause systems to perform incorrectly or even to crash.

7.2. INTERNET TRANSACTIONS AND SECURITY

The ease of shopping and comparing products and prices online has made it an attractive option for many shoppers. How can we make sure our transactions are safe and our credit card information going only where we intend it to ? There are several ways to help ensure safe transactions on the Internet, and more are becoming possible all the time. Some of these include :

1. Stored-value cards (cards that we can buy with specified, loaded Rs. amounts)
2. Smart cards (cards that can act as credit cards, debit cards and/or stored-value cards)
3. Point-of-sale (POS) devices (like our PDA or mobile phone)
4. Digital cash
5. E-wallets
6. Online payment services like PayPal

The most important method for paying for the things we purchase online is still the credit card. The following list provides some tips on how to make sure our transaction is secure and to do so following steps can be taken :

(*a*) **Use the Latest Internet Browser :** The program that we use to surf the Internet is called a browser. This software has built-in encryption capabilities that scramble the information we send to a server. Using the most recent browser ensures that the data is protected using the latest encryption technology. This technology also uses a Secure Sockets Layer (SSL), which is an Internet security protocol used by Internet browsers and Web servers to transmit sensitive information.

(*b*) **Look for Digital Certificates :** That authenticate the entity we are dealing with. Independent services like VeriSign will authenticate the identity of the Web site we are visiting. Web sites that use this service (usually those that sell items or services online) will have the VeriSign logo. By clicking on the logo, you can be assured that the site is legitimate, rather than a clone of the legitimate company set up to collect your personal and financial information.

(*c*) **Read the Privacy Policy :** The information we enter on the Web site should be kept confidential. Make sure we read the company's privacy policy to ensure that our personal information won't be sold to others.

(*d*) **Only Use One Credit Card** for all of your online purchases.

(*e*) Never give out passwords or user ID information online unless we know who we are dealing with **and why they need it.**

(*f*) **Keep Records of All of Our Internet Transactions.** Watch our credit card statement for the charges and make sure they're accurate.

(*g*) **After we have made Purchases Online, Check our E-mail.** Merchants often send confirmation e-mails or other communications about our order.

7.2.1. Types of Identity Theft

Identity theft can enter into many areas of our lives. It involves any instance where a person uses someone else's identification documents or other identifiers in order to impersonate that person for whatever reason. According to a September 2003 survey conducted by the Federal Trade Commission, an estimated 10 million people in the United States found out they were victims of identity theft in the previous year. More appropriately titled identity fraud, our identity might be stolen in order for someone to commit :

1. Financial Fraud : This type of identity theft includes bank fraud, credit card fraud, computer and telecommunications fraud, social program fraud, tax refund fraud, mail fraud, and several more.

2. **Criminal Activities:** This type of identity fraud involves taking on someone else's identity in order to commit a crime, enter a country, get special permits, hide one's own identity, or commit acts of terrorism. These criminal activities can include:

- Computer and cyber crimes
- Organized crime
- Drug trafficking
- Alien smuggling
- Money laundering

7.2.2. Credit Card Fraud

Have we ever eaten at a restaurant, paid with a credit card, and forgotten to get our copy of the credit card receipt ? Did we know that many of these receipts have our credit card number printed right there for anyone to see (and use) ? And, if we have signed them, our signature is also right there for someone to carefully copy. This can lead to the most simple form of identity theft. With this bit of information, some unscrupulous person can be well on his way to making purchases either by phone or on the Internet using our credit card number. We won't know about it until we get our statement (a good reason why we should always study the charges on our credit card statements!). All they have to have, in most cases, is our mailing address, which can be looked up in a phone book or easily found on the Internet.

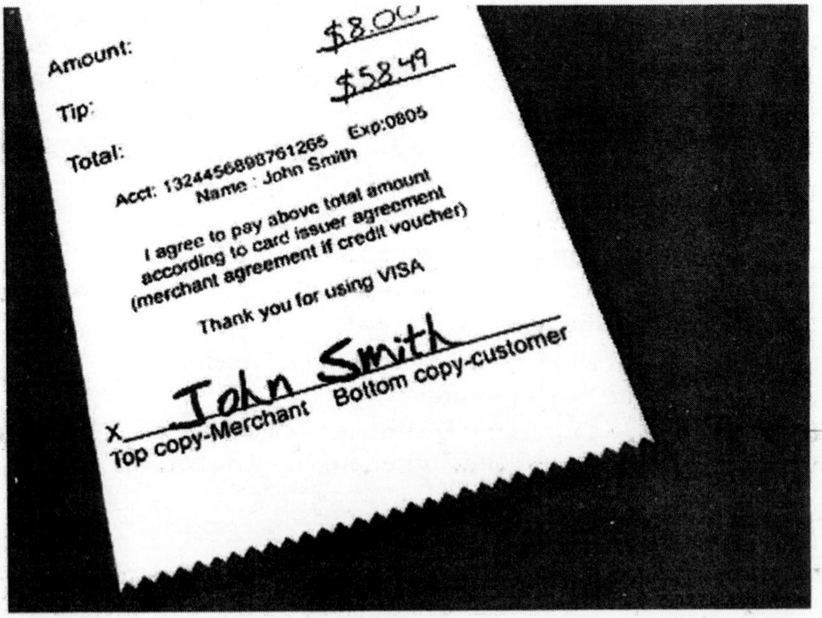

FIGURE 7.1.

Credit card fraud is identity theft in its most simple and common form. It can be accomplished either through a scenario like the one we just mentioned, or it can happen when our pre-approved credit card offers fall into the wrong hands. All a person has to do is get these out of our mailbox (or trash can) and mail them in with a change of address request and start spending. Someone can even apply for a credit card in our name if they have the right information. We won't know a thing about it until the credit card company tracks you down and demands payment for the purchases "you" have a racked up.

7.2.3. How To Protect Ourself

Protecting yourself from identity theft takes proactive effort. We can't simply assume it's not going to happen to us and go on about our life — it can happen to anyone. While you can't ever totally protect ourself from these thieves, you can at least make yourself less attractive as a victim by doing what you can to make it more difficult for them to access our information. Here are some things we can do to protect yourself :

1. **DON'T give out your ID number** unless it is absolutely necessary. Many companies collect more information than they really need. Make sure that it's something they have to have and make sure they'll protect your privacy.
2. **DESTROY any unwanted credit card offers.** This means rip, shred, burn, whatever you can do. These pre-approved offers come almost daily.
3. **DON'T put any other information besides your name and address on your checks,** and keep a close watch on your checkbook both when you're writing checks and when it is lying around. Someone can memorize your name, address and phone number during the short time it takes you to write a check.
4. **SHRED (cross-cut) any sensitive documents** before you throw them into the trash. This may seem like an extreme measure, but dumpster diving happens all the time and turns up a lot more personal information than you may realize.
5. **DON'T carry** your passport, or birth certificate in your wallet or purse. Also, only carry as many credit cards as are absolutely necessary. It has also been suggested that you photocopy everything you carry in your wallet to make canceling things easier in the event that your wallet is stolen.
6. **REVIEW your credit report** every year to make sure there haven't been any new credit cards or other accounts issued (to someone other than you) and to make sure there haven't been inquiries by people you haven't initiated business with.
7. **NEVER give out personal information** on the phone to someone you don't know and who initiated the call. Often, scam artists phone unsuspecting victims pretending to be their financial services company and request information to be provided over the phone. Usually, the story is to "update records" or sell a product. Get their name, phone number and address, and then call them back at the number you have on file or that is printed on the statements you receive.
8. **REVIEW your monthly credit card statement** each month to make sure there aren't any charges showing up that aren't yours. Also, make sure you *get* a monthly statement. If the statement is late, contact the credit card company. You never know when someone may have turned in a change-of-address form so they could make a few more weeks of purchases on your credit card without you noticing.
9. **DON'T mail bills or documents that contain personal data** (like tax forms or checks) from your personal mailbox. Take them directly to the post office or an official postal service mailbox.
10. **REACT QUICKLY if a creditor or merchant calls you about charges you didn't make.** This too may be the first notice you get that someone has stolen your identity. Get as much information from them as you can and investigate immediately.
11. **GUARD deposit slips as closely as you do checks.** Not only do they have your name, address and account number printed on them, but they can also be used to withdraw money from your account. All a thief has to do is write a bad check, deposit it into your account and use the "less cash received" line to withdraw your money.

7.3. ENCRYPTION

When we think of encryption, we probably think of some sort of secret code that prevents others from reading our messages. Whilst this privacy aspect of cryptography is important, it is only one of four aspects that are of particular importance in electronic commerce.

(i) **Authentication** allows customers to be sure that the merchant they are sending their credit card details to is who they say they are. It can also allow merchant's to verify that the customer is the real owner of the credit card being used.

(ii) **Integrity** ensures that the messages have not been tampered with by a third party during transmission.

(iii) **Non-repudiation** prevents customers or merchants denying they ever received or sent a particular message

(iv) **Privacy** prevents third parties from reading intercepted messages

7.3.1. Elements of an Encryption System

The main elements of an encryption system are the plaintext, the cryptographic algorithm, the key and the ciphertext.

(i) The **plaintext** is the raw message or data that is to be encrypted.

(ii) A **cryptographic algorithm or cipher** is a mathematical set of rules that defines how the plain text is to be combined with a key.

(iii) The **key** is a string of digits.

(iv) The **ciphertext** is the encrypted message.

These terms are probably best illustrated through a very simple example.

If we take the phrase "Web store" and add 2 characters to each letter the phrase becomes "ygd uvqtg". Here :

"Web store" is the plaintext

"add x characters to each letter" is the cryptographic algorithm

"2" is the key

"ygd uvqtg" is the ciphertext

There are two main types of encryption in common use today : **secret-key and public-key.**

7.3.2. Secret-Key Encryption

Secret-key encryption, also known as **single-key or symmetric encryption,** involves the use of a single key that is shared by both the sender and the receiver of the message.

After creating the message, the sender encrypts it with their key and passes it to the recipient who then decrypts it by using a copy of the same key used to encrypt it.

A widely used method of secret-key encryption is **the data encryption standard or DES.**

Secret-key encryption does have some limitations, particularly with regard to key distribution. Let us see some of them as follows :

(i) For privacy to be maintained, every transmitter of messages would need to

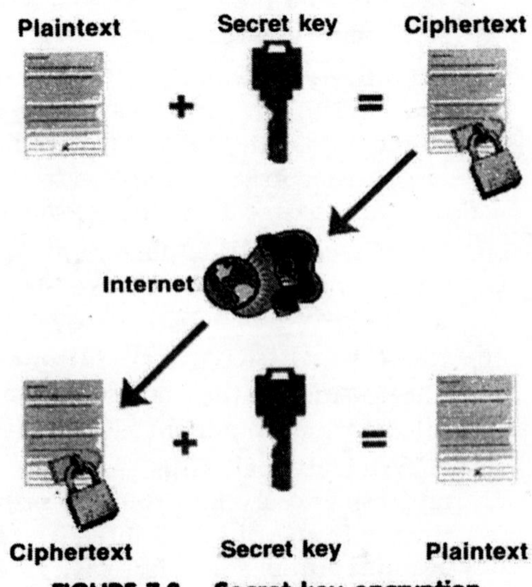

FIGURE 7.2. Secret key encryption.

provide a different key to everyone they intended to communicate with, otherwise every potential recipient would be able to read all messages whether it was intended for them or not.

(*ii*) While this is manageable where a small number of parties are involved (for example, sending a private e-mail to a friend) it is not practical for Web commerce which can involve communicating with thousands of customers.

(*iii*) Another limitation with secret-key encryption is its inability to support non-repudiation. As both parties share the same key it is possible for one party to create a message with the shared secret key and falsely claim the other party had sent it.

(*iv*) **Secret-key encryption on its own is not suitable for Web commerce—instead a system known as public-key encryption is used**

7.3.3. Public-Key Encryption

Public-key encryption, or asymmetric encryption involves the use of two keys, one that can be used to encrypt messages (the public key) and one that can be used to either encrypt them or decrypt them (the private key).

These key pairs can be used in two different ways, to provide privacy or authentication.

Privacy is ensured by encoding a message with the public key as it can only be decoded by the holder of the private key.

FIGURE 7.3.

Authentication is achieved by encoding a message with the private key. Once the recipient has successfully decrypted it with the public key they can be assured it was sent by the holder of the private key.

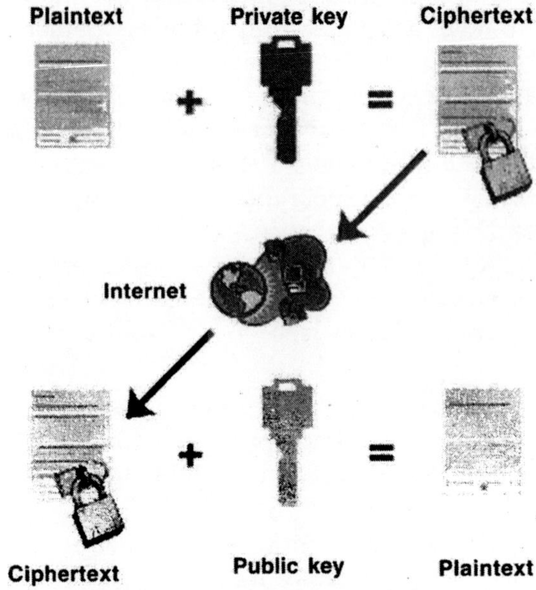

FIGURE 7.4. Public key encryption.

As the public-key can be made widely available (from a server or third party), public-key cryptography does not suffer from the same key distribution and management problems as the secret-key system.

One disadvantage of the public-key system is that it is relatively slow, so when it is being used only for authentication it is not desirable to encrypt the whole message particularly if it is a long one. To get round this a digital signature is used.

7.4. DIGITAL CERTIFICATE

Digital Certificates are the electronic counterparts to driver licenses, passports and membership cards. You can present a Digital Certificate electronically to prove your identity or your right to access information or services online.

Digital Certificates, also known as digital certificates, bind an identity to a pair of electronic keys that can be used to encrypt and sign digital information. A Digital Certificate makes it possible to verify someone's claim that they have the right to use a given key, helping to prevent people from using phony keys to impersonate other users. Used in conjunction with encryption, Digital Certificates provide a more complete security solution, assuring the identity of all parties involved in a transaction.

A Digital Certificate is issued by a Certification Authority (CA) and signed with the CA's private key.

A Digital Certificate typically contains the :

- Owner's public key
- Owner's name
- Expiration date of the public key
- Name of the issuer (the CA that issued the Digital Certificate)
- Serial number of the Digital Certificate
- Digital signature of the issuer

The most widely accepted format for Digital Certificates is defined by the CCITT X.509 international standard; thus certificates can be read or written by any application complying with X.509. Further refinements are found in the PKCS standards and the PEM standard.

7.4.1. Use of Digital Certificates

Digital Certificates can be used for a variety of electronic transactions including e-mail, electronic commerce, groupware and electronic funds transfers. Netscape's popular Enterprise Server requires a Digital Certificate for each secure server.

For example, a customer shopping at an electronic mall run by Netscape's server software requests the Digital Certificate of the server to authenticate the identity of the mall operator and the content provided by the merchant. Without authenticating the server, the shopper should not trust the operator or merchant with sensitive information like a credit card number. The Digital Certificate is instrumental in establishing a secure channel for communicating any sensitive information back to the mall operator.

7.4.2. Need of Digital Certificate

Virtual malls, electronic banking, and other electronic services are becoming more common place, offering the convenience and flexibility of round-the-clock service direct from your home. However, your concerns about privacy and security might be preventing you from taking advantage of this new medium for your personal business. Encryption alone is not enough, as it provides no proof of the identity of the sender of the encrypted information. Without special safeguards, you risk being impersonated online. Digital Certificates address this problem, providing an electronic means of verifying someone's identity. Used in conjunction with encryption, Digital Certificates provide a more complete security solution, assuring the identity of all parties involved in a transaction.

Similarly, a secure server must have its own Digital Certificate to assure users that the server is run by the organisation it claims to be affiliated with and that the content provided is legitimate

7.4.3. Digital Signature

A digital signature functions for electronic documents like a handwritten signature does for printed documents. The signature is an unforgeable piece of data that asserts that a named person wrote or otherwise agreed to the document to which the signature is attached.

A digital signature actually provides a greater degree of security than a handwritten signature. The recipient of a digitally signed message can verify both that the message originated from the person whose signature is attached and that the message has not been altered either intentionally or accidentally since it was signed. Furthermore, secure digital signatures cannot be repudiated; the signer of a document cannot later disown it by claiming the signature was forged.

In other words, Digital Signatures enable "authentication" of digital messages, assuring the recipient of a digital message of both the identity of the sender and the integrity of the message. The requirement form the digital signatures are following :

(*i*) The signature must be a bit pattern that is dependent on the message being signed.

(*ii*) To prevent forgery and denial, the signature must use some information unique to the sender.

(*iii*) The digital signature must be easy to generate.

(*iv*) The storage of a copy of the digital copy must be simple.

(*v*) Forging the signature must be computationally infeasible i.e., either by constructing a fraudulent signature for a given message, or constructing a new message with an existing signature.

(*vi*) The signature must be easy to verify and recognize.

7.4.3.1. Use of Digital Signature for Authentication

Suppose Ramesh wants to send a signed message to Naresh. He creates a message digest by using a hash function on the message. The message digest serves as a "digital fingerprint" of the message; if any part of the message is modified, the hash function returns a different result. Ramesh then encrypts the message digest with her private key. This encrypted message digest is the digital signature for the message.

Ramesh sends both the message and the digital signature to Naresh. When Naresh receives them, he decrypts the signature using Ramesh's public key, thus revealing the message digest. To verify the message, he then hashes the message with the same hash function Ramesh used and compares the result to the message digest he received from Ramesh. If they are exactly equal, Naresh can be confident that the message did indeed come from Ramesh and has not changed since she signed it. If the message digests are not equal, the message either originated elsewhere or was altered after it was signed.

Note that using a digital signature does not encrypt the message itself. If Ramesh wants to ensure the privacy of the message, he must also encrypt it using Naresh's public key. Then only Naresh can read the message by decrypting it with his private key.

It is not feasible for anyone to either find a message that hashes to a given value or to find two messages that hash to the same value. If either were feasible, an intruder could attach a false message onto Ramesh's signature. Specific hash functions have been designed to have the property that finding a match is not feasible, and are therefore considered suitable for use in cryptography.

One or more Digital Certificates can accompany a digital signature. If a Digital Certificate is present, the recipient (or a third party) can check the authenticity of the public key.

7.4.3.2. Validity of Digital Signatures

Normally, a key expires after some period of time, such as one year, and a document signed with an expired key should not be accepted. However, there are many cases where it is necessary for signed documents to be regarded as legally valid for much longer than two years; long-term leases and contracts are examples. By registering the contract with a digital time-stamping service at the time it is signed, the signature can be validated even after the key expires.

If all parties to the contract keep a copy of the time-stamp, each can prove that the contract was signed with valid keys. In fact, the time-stamp can prove the validity of a contract even if one signer's key gets compromised at some point after the contract was signed. **Any digitally signed document can be time-stamped, assuring that the validity of the signature can be verified after the key expires.**

7.4.3.3. Digital Time-Stamping Service

A digital time-stamping service (DTS) issues time-stamps which associate a date and time with a digital document in a cryptographically strong way. The digital time-stamp can be used at a later date to prove that an electronic document existed at the time stated on its time-stamp. For example, a physicist who has a brilliant idea can write about it with a word processor and have the document time-stamped. The time-stamp and document together can later prove that the scientist deserves the Nobel Prize, even though an arch rival may have been the first to publish.

Here's one way such a system could work. Suppose Ramesh signs a document and wants it time-stamped. He computes a message digest of the document using a secure hash function and then sends the message digest (but not the document itself) to the DTS, which sends her in return a digital time-stamp consisting of the message digest, the date and time it was received at the DTS, and the signature of the DTS. Since the message digest does not reveal any information about the content of the document, the DTS cannot eavesdrop on the documents it time-stamps. Later, Ramesh can present the document and time-stamp together to prove when the document was written. A verifier computes the message digest of the document, makes sure it matches the digest in the time-stamp, and then verifies the signature of the DTS on the time-stamp.

The use of a DTS would appear to be extremely important, if not essential, for maintaining the validity of documents over many years. Suppose a landlord and tenant sign a twenty-year lease. The public keys used to sign the lease are set to expire after two years. Solutions such as recertifying the keys or resigning every two years with new keys require the cooperation of both parties several years after the original signing. If one party becomes dissatisfied with the lease, he or she may refuse to cooperate. The solution is to register the lease with the DTS at the time of the original signing; both parties would then receive a copy of the time-stamp, which can be used years later to enforce the integrity of the original lease.

In the future, it is likely that a DTS will be used for everything from long-term corporate contracts to personal diaries and letters. Today, if an historian discovers some lost letters of Gandhi Ji, their authenticity is checked by physical means. But a similar find 100 years from now may consist of an author's computer files; digital time-stamps may be the only way to authenticate the find.

7.4.3.4. Legal Status of Documents Signed with Digital Signatures

If digital signatures are to replace handwritten signatures they must have the same legal status as handwritten signatures, i.e., documents signed with digital signatures must be legally binding. The Australian Government and most states and territories in Australia have already enacted legislation that gives electronic communications the same status as written communications at law (both criminal and civil). These are known as the Electronic Transactions Acts. There are some limitations on when electronic communications are effective, but the basic principle is that transactions are not invalid because they took place electronically.

However, since the validity of documents with digital signatures has never been challenged in court, their legal status is not yet well-defined. Through such challenges, the courts will issue rulings that collectively define which digital signature methods, key sizes, and security precautions are acceptable for a digital signature to be legally binding.

Digital signatures have the potential to possess greater legal authority than handwritten signatures. If a ten page contract is signed by hand on the tenth page, one cannot be sure that the first nine pages have not been altered. However, if the contract was signed with digital signatures, a third party can verify that not one byte of the contract has been altered.

Currently, if two people want to digitally sign a series of contracts, they might first sign a paper contract in which they agree to be bound in the future by any contracts digitally signed by them with a given signature method and minimum key size.

7.5. SECURE SOCKET LAYER (SSL)

Secure Socket Layer (SSL) is a protocol designed and implemented by Netscape Communications. Netscape claims it is designed to work, as the name implies. At the socket layer, to protect any higher level protocol build on sockets, such as Telnet, FTP and HTTP.

Secure Socket Layer (SSL) technology allows web browsers and web servers to communicate over a secure connection. In this secure connection, the data that is being sent is encrypted before being sent and then is decrypted upon receipt and before processing. Both the browser and the server encrypt all traffic before sending any data. SSL addresses the following important security considerations.

(i) ***Authentication*** : During your initial attempt to communicate with a web server over a secure connection, that server will present our web browser with a set of credentials in the form of a server certificate. The purpose of the certificate is to verify that the site is who and what it claims to be. In some cases, the server may request a certificate that the client is who and what it claims to be (which is known as client authentication).

(ii) ***Confidentiality***: When data is being passed between the client and the server on a network, third parties can view and intercept this data. SSL responses are encrypted so that the data cannot be deciphered by the third party and the data remains confidential.

(iii) ***Integrity*** : When data is being passed between the client and the server on a network, third parties can view and intercept this data. SSL helps guarantee that the data will not be modified in transit by that third party.

7.5.1. Use of SSL

To use SSL, an application server must have an associated certificate for each external interface, or IP address, that accepts secure connections. The theory behind this design is that a server should provide some kind of reasonable assurance that its owner is who you think it is, particularly before receiving any sensitive information. It may be useful to think of a certificate as a "digital driver's license" for an Internet address. It states with which company the site is associated, along with some basic contact information about the site owner or administrator.

The digital certificate is cryptographically signed by its owner and is difficult for anyone else to forge. For sites involved in e-commerce or in any other business transaction in which authentication of identity is important, a certificate can be purchased from a well-known certificate authority (CA).

Sometimes authentication is not really a concern—for example, an administrator may simply want to ensure that data being transmitted and received by the server is private and cannot be snooped by anyone eavesdropping on the connection. In such cases, you can save the time and expense involved in obtaining a CA certificate and simply use a self-signed certificate.

SSL uses *public key cryptography*, which is based on *key pairs*. Key pairs contain one public key and one private key. If data is encrypted with one key, it can be decrypted only with the other key of the pair. This property is fundamental to establishing trust and privacy in transactions. For example, using SSL, the server computes a value and encrypts the value using its private key. The encrypted value is called a *digital signature*. The client decrypts the encrypted value using the server's public key and compares the value to its own computed value. If the two values match, the client can trust that the signature is authentic, because only the private key could have been used to produce such a signature.

Digital certificates are used with the HTTPS protocol to authenticate web clients. The HTTPS service of most web servers will not run unless a digital certificate has been installed. Use the procedure outlined later to set up a digital certificate that can be used by your web server to enable SSL.

One tool that can be used to set up a digital certificate is keytool, a key and certificate management utility that ships with the J2SE SDK. It enables users to administer their own public/ private key pairs and associated certificates for use in self-authentication (where the user authenticates himself or herself to other users or services) or data integrity and authentication services, using digital signatures. It also allows users to cache the public keys (in the form of certificates) of their communicating peers.

7.6. AUTHENTICATION

As stated earlier, encryption is the process of taking all of the data that one computer is sending to another and encoding it into a form that only the other computer will be able to decode. Another process, **authentication**, is used to verify that the information comes from a trusted source. Basically, if information is "authentic," we know who created it and we know that it has not been altered in any way since that person created it. These two processes, encryption and authentication, work hand-in-hand to create a secure environment.

There are several ways to authenticate a person or information on a computer:

1. **Password :** The use of a user name and password provides the most common form of authentication. You enter your name and password when prompted by the computer. It checks the pair against a secure file to confirm. If either the name or the password does not match, then you are not allowed further access.

2. **Pass Cards :** These cards can range from a simple card with a magnetic strip, similar to a credit card, to sophisticated smart cards that have an embedded computer chip.

3. **Digital Signatures :** A digital signature is basically a way to ensure that an electronic document (e-mail, spreadsheet, text file) is authentic.

Recently, more sophisticated forms of authentication have begun to show up on home and office computer systems. Most of these new systems use some form of **biometrics** for authentication. Biometrics uses biological information to verify identity. Biometric authentication methods include:

(*a*) Fingerprint scan
(*b*) Retina scan
(*c*) Face scan
(*d*) Voice identification

7.7. INTRODUCTION TO VPN

A Virtual Private Network (VPN) connects the components of one network over another network. VPNs accomplish this by allowing the user to *tunnel* through the Internet or another public network in a manner that provides the same security and features formerly available only in private networks (see Figure 7.5).

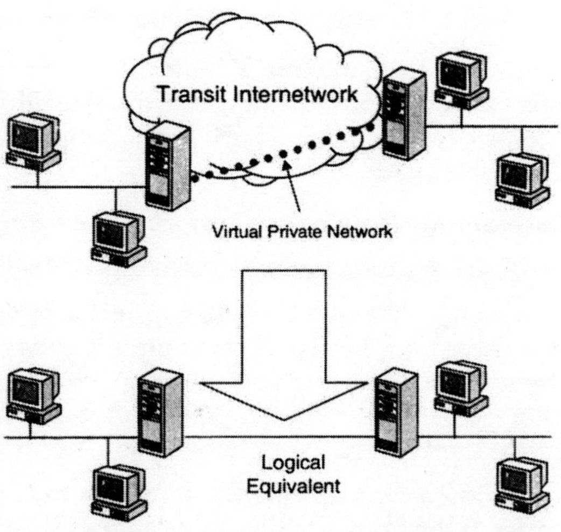

FIGURE 7.5. Virtual private network.

VPNs allow users working at home or on the road to connect in a secure fashion to a remote corporate server using the routing infrastructure provided by a public internetwork (such as the Internet). From the user's perspective, the VPN is a point-to-point connection between the user's computer and a corporate server. The nature of the intermediate internetwork is irrelevant to the user because it appears as if the data is being sent over a dedicated private link.

VPN technology also allows a corporation to connect to branch offices or to other companies over a public internetwork (such as the Internet), while maintaining secure communications. The

VPN connection across the Internet logically operates as a Wide Area Network (WAN) link between the sites.

In both of these cases, the secure connection across the internetwork appears to the user as a private network communication—despite the fact that this communication occurs over a public internetwork—hence the name *Virtual Private Network*.

VPN technology is designed to address issues surrounding the current business trend toward increased telecommuting and widely distributed global operations, where workers must be able to connect to central resources and must be able to communicate with each other.

7.7.1. Common Uses of VPNs

Let us see VPN situations in more detail.

Remote User Access Over the Internet

VPNs provide remote access to corporate resources over the public Internet, while maintaining privacy of information. Figure 7.6 shows a VPN used to connect a remote user to a corporate intranet.

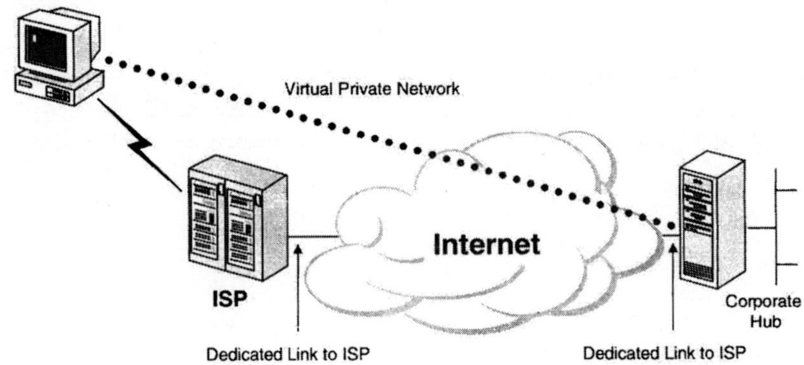

FIGURE 7.6. Using a VPN to connect a remote client to a private LAN.

Rather than making a long distance (or 1-800) call to a corporate or outsourced Network Access Server (NAS), the user calls a local ISP. Using the connection to the local ISP, the VPN software creates a virtual private network between the dial-up user and the corporate VPN server across the Internet.

Connecting Networks Over the Internet

There are two methods for using VPNs to connect local area networks at remote sites:

Using dedicated lines to connect a branch office to a corporate LAN. Rather than using an expensive long-haul dedicated circuit between the branch office and the corporate hub, both the branch office and the corporate hub routers can use a local dedicated circuit and local ISP to connect to the Internet. The VPN software uses the local ISP connections and the Internet to create a virtual private network between the branch office router and corporate hub router.

Using a dial-up line to connect a branch office to a corporate LAN. Rather than having a router at the branch office make a long distance (or 1-800) call to a corporate or outsourced NAS, the router at the branch office can call the local ISP. The VPN software uses the connection to the local ISP to create a VPN between the branch office router and the corporate hub router across the Internet.

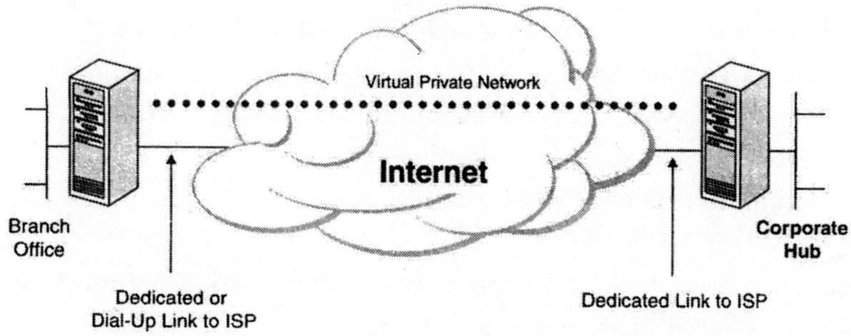

FIGURE 7.7. Using a VPN to connect two remote sites.

In both cases, the facilities that connect the branch office and corporate offices to the Internet are local. The corporate hub router that acts as a VPN server must be connected to a local ISP with a dedicated line. This VPN server must be listening 24 hours a day for incoming VPN traffic.

Connecting Computers Over an Intranet

In some corporate internetworks, the departmental data is so sensitive that the department's LAN is physically disconnected from the rest of the corporate internetwork. Although this protects the department's confidential information, it creates information accessibility problems for those users not physically connected to the separate LAN.

VPNs allow the department's LAN to be physically connected to the corporate internetwork but separated by a VPN server. The VPN server is not acting as a router between the corporate internetwork and the department LAN. A router would connect the two networks, allowing everyone access to the sensitive LAN. By using a VPN, the network administrator can ensure that only those users on the corporate internetwork who have appropriate credentials (based on a need-to-know policy within the company) can establish a VPN with the VPN server and gain access to the protected resources of the department. Additionally, all communication across the VPN can be encrypted for data confidentiality. Those users who do not have the proper credentials cannot view the department LAN.

FIGURE 7.8. Using a VPN to connect two computers on the same LAN

7.7.2. Basic VPN Requirements

Typically, when deploying a remote networking solution, an enterprise needs to facilitate controlled access to corporate resources and information. The solution must allow roaming or remote clients to connect to LAN resources, and the solution must allow remote offices to connect to each other to share resources and information (LAN-to-LAN connections). In addition, the solution must ensure the privacy and integrity of data as it traverses the Internet. The same concerns apply in the case of sensitive data traversing a corporate internetwork.

Therefore, a VPN solution should provide at least all of the following :

1. **User Authentication** : The solution must verify the user's identity and restrict VPN access to authorized users only. It must also provide audit and accounting records to show who accessed what information and when.

2. **Address Management** : The solution must assign a client's address on the private net and ensure that private addresses are kept private.

3. **Data Encryption** : Data carried on the public network must be rendered unreadable to unauthorized clients on the network.

4. **Key Management** : The solution must generate and refresh encryption keys for the client and the server.

5. **Multiprotocol Support** : The solution must handle common protocols used in the public network. These include IP, Internet Packet Exchange (IPX), and so on.

7.7.3. Tunneling Basics

Tunneling is a method of using an internetwork infrastructure to transfer data for one network over another network. The data to be transferred (or *payload*) can be the frames (or packets) of another protocol. Instead of sending a frame as it is produced by the originating node, the tunneling protocol encapsulates the frame in an additional header. The additional header provides routing information so that the encapsulated payload can traverse the intermediate internetwork.

The encapsulated packets are then routed between tunnel endpoints over the internetwork. The logical path through which the encapsulated packets travel through the internetwork is called a *tunnel*. Once the encapsulated frames reach their destination on the internetwork, the frame is unencapsulated and forwarded to its final destination. Tunneling includes this entire process (encapsulation, transmission, and unencapsulation of packets).

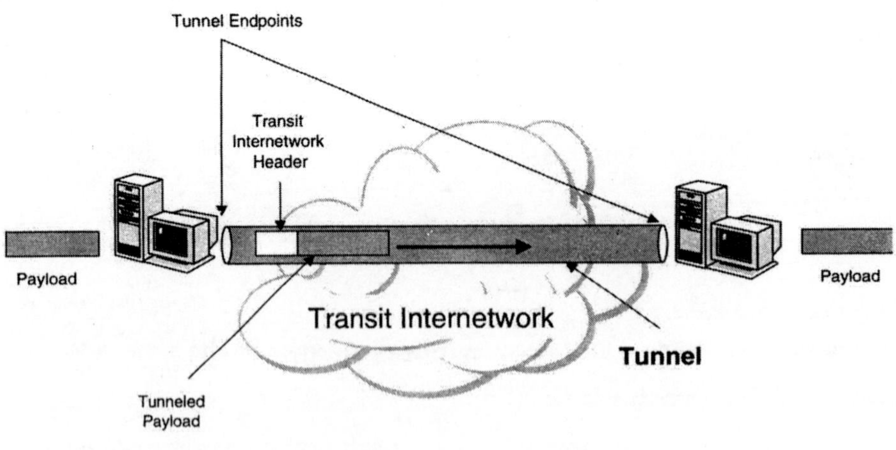

FIGURE 7.9. Tunneling.

The transit internetwork can be any internetwork—the Internet is a public internetwork and is the most widely known real world example. There are many examples of tunnels that are carried over corporate internetworks.

7.7.4. Security Mechanisms are Available Through VPN

Following security mechanisms are available through VPN :

1. Authorisation : VPN connections are only created for users and routers that have been authorised. For Windows 2000, authorization of VPN connections is determined by dial-in properties on the user account and remote access policies. If a user or router is not authorised for such connections, the server will disable them.

2. Authentication : This is a vital security concern. Authentication takes place at two levels:

(*a*) **Machine-Level Authentication** : When IPSec protocol is used for a VPN connection, machine-level authentication is performed through the exchange of machine certificates during the establishment of the IPSec connection.

(*b*) **User-Level Authentication :** Before data can be sent over the PPTP (Point to Point Transfer Protocol) or L2TP (Layer 2 Tunneling Protocol) tunnel, the user must be authenticated. This is done through the use of a PPP authentication method.

3. Data Encryption – the protocols used to create VPN connections allow encrypted data to be sent over a network. Although it is possible to have a non-encrypted connection, this is not recommended. Note that data encryption for VPN connections does not provide end-to-end security (encryption), but only security between the client and the VPN server. In order to provide a secure end-to-end connection, the IPSec protocol can be used once a VPN connection has been established.

4. Packet Filtering – in order to enhance security of the VPN server, packet filtering must be configured so that the server only performs VPN routing. To this end, appropriate RRAS filters should be used (for Windows 2000) on the Internet interface of the VPN.

7.7.5. VPN Management Issues

As we need for secure electronic transmission becomes ever more desirable, corporates are turning quickly to one possible solution : virtual Private Networks (VPN).

But whenever any corporate decides to go for VPN, they think :

(*i*) How will they push for VPN technology.

(*ii*) What will be implementation cost of VPN.

(*iii*) Anything from current infrastructure can be used or not.

(*iv*) Who will private proper training to their staff.

A VPN needs to be viewed as a very dynamic component of the corporation. It is clear that time to time encryption algorithm and tunneling protocols will change. So our VPN should have a software upgrade to support these new changes else more expensive solution, that is hardware upgrade is required. Time to time bandwidth upgradition is required so last factor should be considered. Finally let us discuss some management issues in VPN, that is user database tunneling, key management, software, hardware.

1. User/Database : We define the 'user database' as the total number of users who authenticate to the VPN. It is 'easy' to estimate how much our company will grow in size in say, a year's time frame. We can look at a historical year-by-year growth as well as work with Human Resources Department to find out how many people they have been authorized to hire for the coming year. But what if our company merges or becomes a player in an acquisition ? In many cases the ability to forecast this type of change – and plan accordingly—is near impossible. But if it happens, exactly how many users will be added to our VPN ? The mantra here is to play the 'what if' game and make sure that our proposed solution(s) support growth. Ensure that the user database can scale. Ensure that the authentication servers can also scale—while minimizing performance degradation. Our total VPN solution (user database, authentication device, Internet access, data tunnels, and encryption devices) needs to be able to scale. Ensure that the recommendation report states the possibility of expansion due to mergers and/or acquisitions, and that an action plan is included.

2. Issue of Tunneling : Deploying a VPN with no regard to the tunnels being used can be a costly mistake in terms of network performance and, ultimately, in financial terms.

Tunnel management can be thought of in terms of what tunneling protocol(s) will be used, network subnetting, and bandwidth absorption. It is imperative to plan what protocol(s) will be used in the VPN. There are 'legacy tunnel protocols' (such as PPTP) as well as 'new technology tunnel protocols' (such as IPSec) to chose from. Perhaps we will need both protocols initially. The move in the INFOSEC Industry is towards IPSec.

3. Key Management : Key management can be defined as the generation, distribution, storage, and security of keys. Do we have enough staff on hand to handle each of these tasks ? Who will generate the keys ? Who is the backup to that person if she or he leaves or falls ill ? Probably most *important* here is the distribution of the keys to the VPN users. How will the keys be distributed and at what cost ? Is this method secure ? Have we provided for a means of key recovery in the event of an honest accident or a disgruntled (ex)employee ? The idea here is like the other areas: think and think some more when drafting your VPN solution report. Is there an acceptable risk of key loss (*i.e.*, data loss) ? And if so, what is that risk and who is held responsible ?

4. Software and Hardware Management : VPN Management cannot be complete without thorough investigation into both software and hardware issues. These two areas will play a vital role in the long-term financials of our company. It is important to note that we are not talking about 'normal' software upgrades, that is the installation of the latest VPN software release. Those should be obvious! Instead, we are speaking of the 'hidden' costs of the underlying operating system (if applicable). These costs can be defined as the resources (time, money, staff) it takes to apply patches and upgrades to the servers the VPN software runs on.

Who will keep track of updates and patches to the VPN software and the client software, as well as the server software ? How will the updates be pushed out to remote users ? Who will be responsible for maintaining the security of the VPN servers ? What is the cost if a server is compromised, and how will it affect the VPN ? What is the plan of action when a server is compromised ? Are the resources in place and outlined in the recommendation report ?

5. Support : When putting together our VPN plan, take care to include all areas of support, not just technical support from the solution provider/reseller. Take stock of your current Help Desk team and find out if they can handle another application. Do we have enough manpower ? Can the staff handle the technical problems bound to arise in tunneling (especially IPSec) ? What is the escalation procedure for incoming calls ? Will full documentation be available to all who need it in order to help educate the end user as well as the support staff ? Also make sure that our engineering staff are ready to support the product. Will training be necessary and, if so, at what cost ? Is there a separation of duties among the staff so as to support the Corporate Security Policy ? What sort of monitoring will occur ? How often (daily, weekly, etc.) will the logs be analyzed ? If the logs are backed up to another server (FTP) is that server secure (access control list, encryption, etc.) ?

SUMMARY

In this chapter, we have discussed some basic concerns of web security and transaction security. Encryption is very simple, that is to transmit message with secret code which is known to receiver. We have seen various encryption method like private key and public key encryption. We also discussed some another security mechanisms like digital certificate, SSL and VPN. VPN is used by corporate houses as their personal network.

REVIEW QUESTIONS

1. "Web security is major concern." Justify it and also discuss the some primary security services categories.
2. Now a days most of the e-commerce transactions takes place on internet. What steps we can take to ensure internet transaction security.
3. What do you understand by Encryption and what are various elements of Encryption system ?
4. Differentiate between Secret-key encryption and public-key encryption ?
5. What is the use of digital certification e-commerce ?
6. Define SSL and how it becomes functional.
7. "VPN can fulfill the need of corporate organization about secure treansmission." Justify the statement.

❑❑❑

8

CHAPTER

ELECTRONIC PAYMENT SYSTEMS

8.1. INTRODUCTION

INSIDE THIS CHAPTER

8.1. Introduction
8.2. Secure Electronic Transaction (SET) Protocol
8.3. Payment Gateway
8.4. Types of Elecironic payment System
8.5. Online Credit Card Based System
8.6. Debit Cards Based EPS
8.7. Electronic Fund Transfer
8.8. Electronic Commerce and Banking
8.9. Electronic Commerce and Online Publishing
8.10. Electronic Commerce and Retailing
8.11. Online Shopping

Electronic Payment Systems are becoming central to online business process innovation as company look for ways to serve customers faster and at lower cost. Emerging innovation in the payment for goods and services in electronic commerce promise to offer a wide range of new business opportunities.

Electronic Payment Systems and E-commerce are linked given that online consumers must pay for products and services.

Electronic payment systems are non-credit-card online payment systems. The goal of their development is to create analogs of checks and cash on the Internet, i.e. to implement all or some of the following features:

1. Protecting customers from merchant's fraud by keeping credit card numbers unknown to merchants.
2. Allowing people without credit cards to engage in online transactions.
3. Protecting confidentiality of customers.
4. In some cases providing anonymity of customers ("electronic cash").

The problems in implementing electronic payment systems, especially anonymous electronic money, are :

(*i*) Preventing double spending: copying the "money" and spending it several times. This is especially hard to do with anonymous money.

(*ii*) Making sure that neither the customer nor the merchant can make an unauthorized transaction.

(*iii*) Preserving customer's confidentiality without allowing customer's fraud.

While electronic payment systems have not gained a very wide popularity, except for PayPal system used on online auctions, such as eBay, they may become more popular in the future if more businesses start using them. Electronic payment systems may be more convenient for international online business due to differences in credit card customer protection laws in different countries.

8.1.1. Problems with Traditional Payment Methods

Traditional payment methods do not work online for the following reasons :

(*i*) **Lack of Convenience** : Traditional payment methods generally require that the consumer leave the online platform and use the telephone or send a check in order to make payment.

(*ii*) **Lack of Security** : In order to make a traditional payment over the internet, a consumer has to provide card/payment account details and other personal information online. Leaving the internet and providing the card/payment account details over the telephone and/or by mail also increase the security risks.

(*iii*) **Lack of Coverage** : Credit cards only work with signed-up merchants and do not generally support individual-to-individual or direct business-to-business payment transactions.

(*iv*) **Lack of Eligibility** : Not all potential buyers have suitable credit ratings to allow them access to credit cards and/or checking accounts.

(*v*) **Lack of Support for Micro Transaction** : Many payments made over the Internet are of sufficiently low value that the cost of a phone call or letter may be too high of an overhead.

8.2. SECURE ELECTRONIC TRANSACTION (SET) PROTOCOL

The growth of the Internet over the past few years has been explosive. Thus, today a surfer can purchase a variety of goods on the Internet, as opposed to merely accessing information. However, what is impeding rapid growth of this business is the consumer perception of poor security on the Net. Most payment methods revolve around the credit card, and consumers remain hesitant to reveal their card information on the Internet. Also, not surprisingly, credit card frauds on the Internet have registered a dramatic increase.

To address these growing security concerns and pave the way for uninhibited growth of electronic commerce on the net, the two leading credit card brands, Visa and MasterCard, teamed up some years ago to develop a common standard to process card transactions on the Internet, called the Secure Electronic Transaction (SET) standard. As the name implies, SET is a standard which will ensure that credit card and associated payment order information travels safely and securely between the various involved parties on the Internet.

8.2.1. A Sample SET Session

Before getting into details of SET, we shall take a simple example to describe how SET works from the consumer's perspective.

1. The consumer accesses the merchant's web site, goes through the various goods on display and selects what he or she wants. Perhaps there is a virtual shopping cart where he or she drops all the items to be purchased. At the end, the customer proceeds to the virtual checkout counter. A screen pops up giving details, including the cost of all the items the shopper is purchasing, plus taxes and shipping costs.
2. Then the screen asks for the payment method and the consumer chooses to pay through a credit card using SET.

3. Immediately, a special software on the consumer's PC called Digital Wallet is invoked, and it asks the customer to choose one credit card from the many he or she possesses.
4. The consumer chooses a card, and the electronic transaction using SET is underway. A few seconds later, there is a confirmation that this order has been processed.

8.2.2. Objectives of SET

SET addresses seven major business requirements :

1. Provide confidentiality of payment information and enable confidentiality of order information that is transmitted along with the payment information.
2. Ensure the integrity of all transmitted data.
3. Provide authentication that a cardholder is a legitimate user of a branded payment card account.
4. Provide authentication that a merchant can accept branded payment card transactions through its relationship with an acquiring financial institution.
5. Ensure the use of the best security practices and system design techniques to protect all legitimate parties in an electronic commerce transaction.
6. Create a protocol that neither depends on transport security mechanisms nor prevents their use.
7. Facilitate and encourage interoperability among software and network providers.

Point 1 ensures that unauthorized parties cannot view card information. Point 2 ensures that the information cannot be changed or tampered. Points 3 and 4 ensure that the cardholder and merchant are really who they claim they are. Hence, in essence, this framework, if implemented effectively, will allow both buyers and sellers to transact in total confidence in an open network.

8.2.3. Mechanics of SET

What happens in an actual SET transaction is a bit more involved. The method has been developed keeping in mind the basic objectives of SET: confidentiality, integrity and authentication of both sender and receiver.

To ensure integrity, we use a one-way hashing algorithm on the message to generate a message digest. This algorithm uses statistical methods to compute a checksum (or message digest) from the characters in the message. If the content or position of even a single character is changed, the message digest will not match. To secure the message digest itself from tampering, the digest is encrypted using the sender's private key. The encrypted form of the message digest is called the digital signature of the sender. The receiver can decrypt the digest using the sender's public key, regenerate the digest and compare. This also ensures that the message has really come from the sender (nobody else knows the sender's private key) and hence accomplishes the objective of authenticating the sender.

To ensure confidentiality, the entire message is encrypted. Since the message could be large, we use the more efficient symmetric encryption. A random symmetric key is generated and used to encrypt the message. To secure the symmetric key itself, it is encrypted using the public key of the receiver.

Finally, the symmetrically encrypted message and the random symmetric key encrypted using the receiver's public key are sent to the receiver. Only the receiver can decrypt the symmetric key, since only he possesses the private key of the receiver. Thus we have achieved authentication of the receiver.

The receiver uses his private key to decrypt the random symmetric key. Then he uses the random symmetric key to decrypt the main message. He locates the sender's digital signature and using the public sender's key, he decrypts it and retrieves the message digest. He regenerates the message digest using the known hashing function and compares it to the retrieved digest. If they match, he is assured that the message has not been tampered with and has indeed originated from the sender.

One last thing needs to be covered: the concepts of digital certificates and certifying authorities. Before two parties start communicating, each wants to be sure about the authenticity of each other's public keys. For instance how is the receiver sure that the public key of the sender is what is mentioned in the message and not something else? To resolve this, we use trusted third parties, also called certifying authorities (CAs). A certifying authority could be a financial institution, a government body, an international body or a card brand (such as Visa or MasterCard). A certifying authority would create a digital document containing the entity's details and public key and then sign it digitally (this involves creating a message digest and encrypting it under the CA's private key). This digital document is called a digital certificate. A good analogy for this is when we take a birth certificate (public key) to the consulate (certifying authority) to get it attested (digital signature) to yield an attested birth certificate (digital certificate). Since people trust the consulate (CA), they will, in turn, trust the attested birth certificate (digital certificate).

What if the sender or receiver does not know of the existence of (does not trust) the certifying authority itself? This can very often happen if the two parties are in different countries. To overcome this, SET mandates the establishment of a hierarchy of trust. This means that the CA itself will be trusted by a parent CA, which in turn will be trusted by another CA until it goes to a root CA which is trusted by everybody. For example, a financial institution in UAE can be trusted by the regional office of the Visa or MasterCard, which in turn is trusted by the Visa or MasterCard international office, which in turn can be trusted by a UN body. The UN body can function as the root CA who everybody trusts. Thus each party vouchsafes for the level below, by issuing digital certificates. The receiver will authenticate the sender by traversing this hierarchy of trust all the way to the root CA. In an actual implementation, this entire process should only take a few seconds, because it is completely automated.

8.2.4. An SET Implementation

Before we apply this to an actual implementation of SET, let us first examine how a credit card transaction is processed in the physical non-SET world.

(*i*) The cardholder presents the card to the merchant, who in turn swipes the card on a POS terminal.

(*ii*) An electronic message containing the card number and amount is then sent to the acquiring bank, with whom the merchant is associated.

(*iii*) The acquiring bank, in turn, forwards the message to the card brand's (Visa or MasterCard) central computer.

(*iv*) The card brand forwards the message to the issuing bank, which initially issued the card to the cardholder.

(*v*) The issuing bank checks the available credit limit on the card and sends back an authorization.

(*vi*) This authorization travels all the way back, until it reaches the merchant's POS terminal.

Now we will study the detail mechanism of set implementation as follows :

(*a*) The cardholder goes to a merchant's web site and selects the items he or she wants to purchase. The cardholder then clicks on the virtual checkout button or its equivalent.

(*b*) This triggers wallet software to be invoked on the cardholder's PC. The software presents several credit cards which the cardholder possesses, and one is chosen. The wallet software also receives the digital certificates of two entities: the merchant and the acquiring bank (also called a payment gateway). These two certificates are validated by traversing the hierarchy of trust, through messages sent on the Internet to all the entities on the trust chain.

(*c*) The wallet software then generates a message containing two parts: the order information and the payment information. The order information contains information confirming the order, whereas the payment information contains the card number and the amount. The payment information is encrypted using a random symmetric key, which, in turn, is encrypted with the payment gateway's public key, so that only the payment gateway can decrypt it. In other words, the merchant will never know the details of the card number of its customer. This data is sent automatically to the merchant's web site.

(*d*) The merchant's computer will first validate the cardholder's digital certificate. Then it will send the payment information to the payment gateway (which is the acquiring bank's computer).

(*e*) The payment gateway will verify the digital certificates of both the merchant and the card holder and decrypt the message to access the card number and the amount.

(*f*) Then the payment gateway will interface with the legacy systems of the acquiring bank to send the transaction to the card brand, which will then send it to the issuing bank for authorization.

(*g*) This authorization response is then encrypted in the usual fashion and sent to the merchant, who, in turn, will validate the message and store the response. Then the merchant will arrange to ship the goods.

All of these transactions happen on the Internet and are quite transparent to the cardholder.

8.2.5. Practical Requirements for SET

SET has been introduced and is working successfully in several countries. The prerequisites for SET are :

(*i*) The issuing bank will need to request that its **cardholders acquire digital certificates.** Typically, it will supply a wallet software, which will set up a digital certificate for the cardholder automatically.

(*ii*) There has to be a reasonable number of **merchant web sites which support SET.** For this to happen, the merchants' acquiring banks have to implement payment gateways. Payment gateway software is now available from several certified leading software vendors. Currently the number of SET-enabled web sites is small; however, this is expected to grow substantially in the near future.

8.3. PAYMENT GATEWAY

A payment gateway is an e-commerce application service provider service that authorizes payments for e-businesses, online retailers, bricks and clicks, or traditional brick and mortar. It is the equivalent of a physical POS(Point-of-sale) terminal located in most retail outlets. Payment gateways encrypt sensitive information, such as credit card numbers, to ensure that information passes securely between the customer and the merchant.

8.3.1. How Payment Gateways Work

A payment gateway facilitates the transfer of information between a payment portal (such as a website) and the Front End Processor or acquiring bank; quickly and securely.

When a customer orders a product from a payment gateway enabled merchant, the payment gateway performs a variety of tasks to process the transaction; completely invisible to the customer. For example :

(*i*) A customer places order on website by pressing the 'Submit Order' or equivalent button, or perhaps they enter their card details using an automatic phone answering service.

(*ii*) If the order is via a website, the customer's web browser encrypts the information to be sent between their browser and the merchant's webserver. This is usually done via SSL (Secure Socket Layer) encryption.

(*iii*) The merchant then forwards the transaction details through to their payment gateway which holds the detail of their merchant account transaction. This is often another SSL encrypted connection to the payment server hosted by the payment gateway.

(*iv*) The payment gateway which receives the transaction information from the merchant forwards it to the merchant's acquiring bank.

(*v*) The acquiring bank then forwards the transaction information to the issuing bank (the bank that issued the credit card to the customer) for authorization.

(*vi*) The card issuing bank receives the authorization request and sends a response back to the payment gateway (via the acquiring bank) with a response code. In addition to determining the fate of the payment, (*i.e.* approved or declined) the response code is used to define the reason why the transaction failed (such as insufficient funds, or bank link not available).

(*vii*) The payment gateway receives the response, and forwards it on to the website (or whatever interface was used to process the payment) where it is interpreted and a relevant response then relayed back to the customer.

(*viii*) The entire process typically takes 3-4 seconds.

(*ix*) At the end of the bank-day (or settlement period) the acquiring bank deposits the total of the approved funds in to the merchant's nominated account. This could be an account with the acquiring bank if the merchant does their banking with the same bank, or a scrape account with another bank.

8.3.2. Common Payment Gateways

Let us see some of the payment gatways avialable in market as follows :

1. AssureBuy
2. Authorize.net
3. BluePay
4. Bucks Net Services
5. Chase Paymentech (Orbital Gateway)
6. Creditcall
7. Ctel Technologies
8. Cybersource
9. eKashu
10. eMatters
11. First Atlantic Commerce

12. iTransact
13. LinkPoint (owned by FirstData)
14. Network Merchants
15. Payment Express
16. Paypal
17. Plug 'n Pay
18. ProPay
19. PSBill
20. PayJunction.com
21. SafeMerchant Payment Gateway
22. Skipjack Financial Services
23. Soeasypay
24. Shift4
25. VaultLine
26. Verepay Technologies
27. VeriSign Payflow (Owned by PayPal)
28. Viaklix (owned by USBank)

8.4. TYPES OF ELECTRONIC PAYMENT SYSTEM

Electronic Payment Systems are proliferating following :

(*i*) Banking,
(*ii*) Retail,
(*iii*) Health care,
(*iv*) On-line markets
(*v*) Government
(*vi*) In fact, anywhere money needs to change hands.

Research into electronic payment systems for consumers can be traced back to the 1940s and the first applications—credit cards—appeared soon after. In the early 1970s, the emerging electronic payment technology was labeled electronic fund transfer (EFT). EFT is defined as "any transfer of funds initiated through an electronic terminal, telephonic instrument or computer or magnetic tape so as to order, instruct or authorize a financial institution to debit or credit an account, "EFT utilizes computer and telecommunication components both to supply and to transfer money or financial assets. Transfer is information-based and intangible.

EFT can be segmented into three broad categories :

1. **Banking and Financial Payments :**
   - Large-scale wholesale payments (e.g. bank-to-bank transfer)
   - Small-scale or retail payment (e.g. automated teller machines and cash dispensers)
   - Home banking (e.g. bill payment).

2. **Retailing Payments :**
   - Credit cards (e.g. VISA or Master Card)
   - Private label credit/debit cards
   - Charge Cards (e.g. American Express)

3. **Online electronic commerce payments :**

- Token based payment systems :
  Electronic Cash (e.g. Digital Cash)
  Electronic Checks (e.g. Netcheque)
  Smart Cards or Debit cards (e.g. Mondex Electronic Currency Card)
- Credit Card-based payment systems :
  Encrypted credit cards (e.g. World Wide Web form-based encryption)
  Third party authorization numbers (e.g. First Virtual)

8.4.1. Digital Token-based Electronic Payment Systems

None of the banking or retailing payment methods are completely adequate in their presence form for the consumer-oriented e-commerce environment.

Entirely new forms of financial instrument are also being developed." One such new financial instrument is "electronic tokens" in the form of electronic cash/money or checks".

Electronic tokens are designed as electronic analog of various form of payment backed by a bank or financial institution. Simply stated, electronic tokens are equivalent to cash that is backed by the bank.

Electronic tokens are of three types :

1. **Cash or Real Time :** Transactions are settled with the exchange of electronic currency. An example of on-line currency exchange is electronic can (e-cash)

2. **Debit or Prepaid :** Users pay in advance for the privilege of getting information. Examples of prepaid payment mechanism are stored in smart cards and electronic purses that store electronic memory.

3. **Credit or Postpaid :** The server authenticates that costumer and verifies with the bank that funds are adequate before purchase. Example of postpaid mechanism are credit/debit cards and electronic checks.

8.4.2. Using the Digital Currency

Once the token are purchased, the digital cash software on the customer's PC stores digital memory undersigned by a bank. The user can spend digital money at any shop accepting digital cash, without having to open an account or having to transmit credit card numbers. As soon as the customer wants to make a payment, the software collects and transfers the necessary amount from the stored tokens.

Two types of payment transactions are possible :

1. **Bilateral :** Transaction involving cash are bilateral or two-party in which the merchant checks the authenticity of the note's digital signature by using the bank's public key. If satisfy with the payment, the merchant stores the digital currency on her machine and later deposits it in the bank to cash the face value of the note.

2. **Trilateral :** Transactions involving financial instrument other than cash are usually trilateral or three party (buyer, seller or bank) transactions, in which the "notes" are sent to the merchant, who immediately sends them directly to the digital bank. The bank verifies the validity of these "notes" and ensures that they have not been spent before. The bank then credits the merchant's account.

To detect double spending, banks digital cash is identifiable by serial numbers. In order to be able to detect double-spending "notes" have to be registered in some form so that all "notes" used globally can be uniquely identified. However, it is difficult to match notes with a central registry online. For banks, which handle high volume of micro payments, this verification method is cumbersome and time consuming. In addition, the need of monitor notes forces bank to carry extra overhead because of constant checking and security legs.

8.5. ONLINE CREDIT CARD BASED SYSTEM

Most of online purchases are paid for by a credit card. Merchants like credit card payments because an instant authorization guarantees that the card is valid (as opposed to a check which may bounce). Customers like paying by credit cards because they can easily cancel a transaction in case when they don't receive products or services according to the agreement in the transaction.

While some of credit card payments for online services are performed by phone, most of such payments are made by filling in an online form.

Credit card information submitted by the customer is sent to the bank which has issued the credit card to verify. If the transaction is approved, the merchant notifies the customer that the order has been placed. The actual transfer of money from the credit card bank to the merchant may happen in a few hours, or even in a few days.

Merchants who accept credit card payments pay fee (between 1 and 7 percent of the card charge) for each card charge. In addition, in some cases merchants pay authorization fee for each credit card authorization attempt, as well as other fees related to credit card processing.

In case when a customer is not satisfied with the product or a service, or for other reasons, merchants may issue a refund or a charge-back to the customer's account.

Credit Card payment negotiation involves two steps :

(*i*) The merchant present the customer with product/service price, order confirmation and status, delivery notifications and payment option accepted by the merchant, and the buyer presents the merchant with payment choice and associated information in a secure manner. As of yet, there is no standard way of sending secure payments instructions over the web. Currently, consumers can "shop" – took at content and read product descriptions – in the web environment, but have to go offline in order to use their credit cards to actually make their purchases.

(*ii*) The different payment schemes require consumer to set up accounts and/or buy or download and install special software for their personal computer. However, not all banks can handle different payment systems. In order to avoid losing customer by selecting one payment method over another, some merchant sidestep the confusion caused by multiple payment standards by verifying credit card transactions manually. They take credit card numbers over the Internet, and then at the end of a day, batch the verification process. If there is a problem, they send e-mail to the customers informing them of the problem.

8.5.1. Types of Credit Card Payments

Credit card-based payment can be divided into three categories :

(*i*) **Payments using Plain Credit Card Details :** The easier method of credit card payment is the exchange of unencrypted credit card over a public network such as telephone lines or the Internet. The low level of security inherent in the design of Internet makes this method problematic (any hacker can read a credit card number, and there are programs that scan the Internet traffic for credit card numbers and send the numbers to their programmers). Authentication is also a significant problem and the vendor is usually responsible for ensuring that the person using the credit card is its owner.

(*ii*) **Payments using Encrypted Credit Card Details :** Even if credit card details are encrypted before they are sent over the Internet, there are still certain factors to consider before sending them out. One such factor is the cost of a credit card transaction itself, which might prohibited low-value payments (micropayments).

(*iii*) **Payments using the Third Party Verification :** One solution to security and verification problem is the introduction of a third party to collect and approve payments from one client to another.

8.5.2. Technical Issues

There are several technical issues involved in online credit card payments, below we discuss some of them.

(*i*) **Quick Check for Types :** Since the merchant may be charged for each credit card authorization, it is convenient to check that the credit card number makes sense before sending it to the issuing bank to authorize. There is an easy algorithm to verify a credit card number: the last digit of the credit card number is computed from the other digits using a simple procedure.

(*ii*) **Authenticating the User and Protection from Customer Fraud :** Since the card is not physically present during the transaction, it is practically impossible for a merchant to distinguish a legitimate credit card user from a thief. In online transactions the user is usually asked to provide additional information, such as their address and phone number, and the card's billing address, if different from the customer's address. However, this information can be easily mistyped. While in a telephone transaction an operator can use their judgment to approve or reject a transaction based on how much of the information has matched and how confident the customer sounds.

Another way of verifying a card number is to ask the user to provide the additional digits on the card (the digits which do not appear on the magnetic strip or on a carbon paper when the print of the card is taken). However, online customers may be reluctant to provide this information because of fear of merchant's fraud.

(*iii*) **Protecting Card Numbers in Transmission :** Since information transmitted in an online transaction is sufficient for approval of a credit card charge, it is essential that this information is protected from intruders. The most common way of doing it is to encrypt data in transmission. This is done via SSL.

(*iv*) **Protecting Card Numbers on the Merchant's Site :** In practice, the main vulnerability of dealing with credit card numbers is not the transmission, but the storage. Security experts agree that storing credit card numbers at the merchant's site is a risky practice, and should be avoided. If credit card numbers need to be stored, they should be stored on a secure machine, and preferably in an encrypted form. They should not be stored in a database which is (at least partially) accessible to customers, nor should they be stored (in any form) on the web server. It is the merchant's responsibility to protect customer's information from fraud. An e-commerce website may suffer large losses, including those caused by the loss of customer's trust, it fails to protect confidential customer information.

(*v*) **Protecting from Merchant Fraud :** The other side of protecting a merchant from a customer's fraud is protection of a customer against a merchant's fraud. If the merchant knows enough of the customer's credit card information to be able to authorize a transaction, then the merchant (including many of the merchant's employees) know enough to be able to use the credit card themselves! In the majority of cases the highest priority of the merchant is to protect the reputation of the business and their own, and a fraud is not in the merchant's interests. However, there may be exceptions, such as a desperate owner whose business is about to go broke, a disgruntled employee, or an online scam which uses a fake online business as a coverup for collecting credit card information.

An online customer has to be careful not to be a victim of a merchant's fraud. Using SSL to verify the site's name is a way to avoid sites that pretend to be a part of a respected business, but in fact are not. To verify that a business is legitimate, a customer may try calling the phone number or sending an e-mail. It is always important to check carefully the credit card statement and immediately investigate an unauthorized charge (and possibly cancel the credit card if theft is suspected).

However, one has to check carefully the software that implements this feature, because poorly written code for redirection may expose the merchant's ID or allow the customer to change the amount of the transaction in the request.

8.5.3. Virtual PIN

Virtual PIN, started in 1994 by a company called **First Virtual Holding**, was a system for making credit card payments over the Internet without exposing the credit card number to the merchant. It required no special software for a customer to make a purchase. Virtual PIN relied on difficulty of intercepting and forging e-mail.

To enroll, a customer gives their credit card information and their e-mail address to the First Virtual (this was done by phone). After the credit card information has been verified, the customer receives their PIN by e-mail.

The procedure for purchasing an item using Virtual PIN is as follows :

Step 1 : The customer gives the merchant their Virtual PIN.

Step 2 : The merchant sends the Virtual PIN and the amount of transaction to First Virtual.

Step 3 : First Virtual sends an e-mail to the customer asking to confirm the purchase.

Step 4 : The customer answered "Yes", "No", or "Fraud".

Step 5 : If the answer is "Yes", the merchant is informed that the charge has been accepted. If "No", the charge is declined. If the answer is "Fraud", the charge is investigated.

Even though no encryption was involved, an intruders could not use a virtual PIN without being able to intercept and answer the e-mail message to confirm the purchase.

Unlike credit cards which carry the customer's name, Virtual PIN provided a customer's anonymity from the merchant. The e-mail confirmation of the transaction served as a protection against merchant's fraud.

Unfortunately, while the system has been created for all kinds of online business, the main use of Virtual PIN at the time was for buying and selling pornography. Virtual PIN tried to disassociate itself from this market. Eventually the company abandoned the Virtual PIN and became specialized in sending promotional e-mail.

8.5.4. DigiCash (or E-cash)

DigiCash (also known as E-cash) is an electronic payment system developed by Dr. David Chaum, who is widely regarded as an inventor of digital cash. The system was based on digital tokens called *digital coins*. DigiCash operated as follows :

Step 1 : A customer establishes an account with the bank or other organization that could mint and receive digital coins. The customer's account was backed by real money in some form, for instance it could be linked to the customer's checking account.

Step 2 : The customer also needs to download and install a software called *electronic wallet*.

Step 3 : To obtain DigiCash, the customer uses the electronic wallet to create digital coins. The coins are sent to the bank to sign. When the coins are signed, the equivalent amount of money is withdrawn from the customer's account.

In the proposed protocol the customer also had an option of "blinding" the coins. To blind a coin, the customer multiplies it by a random number *r* before sending it to the bank to sign. The bank signs the data. After the data and its digital signature are sent to the customer, the customer computes the digital signature of the original (non-multiplied) coin by dividing the bank's signature by *r*. This way the bank doesn't know the coin, but the customer, who knows *r*, can trace his/her payments. Blind signatures have not been implemented.

Step 4 : When the customer wants to make a purchase, he/she sends signed digital coins to the merchant. The merchant verifies the bank's signature and deposits the coins to the bank, where they are credited to the merchant's account.

The DigiCash (or E-cash), produced by the company DigiCash BV based in Amsterdam, has never created a market. The company eventually declared bankruptcy. However, the algorithms used in DigiCash are considered fundamental in development of digital money.

8.5.5. CyberCash/CyberCoin

CyberCash is a system that allows customers to pay by a credit card without revealing the credit card number to the merchant. To achieve this, a credit card number is sent to the merchant in an encrypted form.

To enroll, a customer installs a software called *CyberCash wallet* on their computer. At the time of the instalment the wallet generated a pair of a public and a private key. The wallet was protected by a passphrase, and a backup key was stored encrypted on a floppy disk. A CyberCash account was linked to the customer's credit card. A variation of this scheme called *CyberCoin* was linked to the customer's checking account.

A purchase was conducted the following way :

(*i*) When the purchase was initiated, the CyberCash wallet displayed the amount, the merchant's name, and other information. After the customer approved the transaction, an encrypted payment order was sent to the merchant.

(*ii*) The merchant could decrypt some of the information in the order, such as the product list, the address, etc., but not the other (such as the credit card information). The merchant's software would add its own payment information to the order, digitally sign it, and then send it to the CyberCash gateway.

(*iii*) The CyberCash gateway would decrypt the information. The order would be checked for duplicate requests. The gateway would verify that the customer's and the merchant's order information match (i.e. no fraud was committed on either side). Then it would perform the money transfer and send the approval message to the merchant.

The main point of this scheme was to prevent merchant's fraud, and thus allow customers to do business with more merchants without fear of scam. However, CyberCash and CyberCoin were not able to find the market. The main reasons for the failure were the large size of customer's software and the fact that very few merchants would accept CyberCash payment. The company was eventually bought by VeriSign.

8.5.6. PayPal

PayPal is an electronic payment system which can transfer money between its accounts. In order to use PayPal, one has to obtain a PayPal account, which is associated either with the customer's credit card or with their regular bank account. The validity of a credit card is checked by the usual ways. The validity of a checking account is checked as follows: the customer gives PayPal their account number, PayPal makes two small-amount (less than $1) deposits to the account. If the customer is able to tell PayPal the value of these deposits, then the customer is assumed to be a legitimate user of the account.

PayPal provides easy interface to send money to anyone by giving the person's e-mail account. In order for the person to retrieve the money, they must have a PayPal account. To avoid fraud, PayPal sends an e-mail message to both the initiator and the recipient of the transaction.

PayPal is used to settle online auctions, such as eBay auctions. The ease of use and the fact that no credit card is required to use it makes PayPal increasingly popular.

8.5.7. Smart Cards

Smart cards also called *stored value cards*, use magnetic strip technology or integrated circuit chips to store customer-specific information, including electronic money. The cards can be used to purchase goods or services, store information, control access to accounts and perform many other functions.

Unlike magnetic strip cards, which can be read by any magnetic reader, and are therefore vulnerable to loss or theft, a smart card can be password-protected to guarantee that the owner only uses it.

Smart cards can run RSA encryption and can be programmed to generate a pair of public/ private keys. The public key is made publicly readable, but the private key is be stored on the card without anyone being able to copy it. Therefore, to use the private key, the user must physically possess the card.

Smart Card Offer Clear Benefit to Both Merchants and Customers. They reduce cash handling expenses and losses caused by fraud, expedite customer transactions at the checkout counter and enhance customer convenience and safety. In addition many state and federal governments are considering stored value cards such as an efficient option for dispersing government entitlements. Other private sector institutions market stored value products to transit riders, university students, telephone customers, vendoring customers and retail customers.

One successful use of stored value cards is by New York's Metropolitan Transportation Authority (MTA). The MTA is the largest transportation agency in the United States and through its subsidiaries affiliates, operate the New York City subway and public bus system, the long Island Railroad ad Metro North commuter rail system and nine tolled intrastate bridges and tunnels. These facilities serve four million customers each working day. In 1994, the MTA began the operation of an automated fare collection system based on a plastic card with a magnetic strip. The Metro Card is either swiped through a card reader at subway stations or dipped into a farebox on buses where the fare is decremented. All 3,600 MTA buses become operational in 1996. the full complement of 467 subway station is expected to be operational by mid-1997. by 1999, the MTA anticipates more than 1.2 billion electronic fare collection transactions a year on subway and bus lines.

8.5.8. Electronic Checks

Electronic checks are designed to accommodate the many individuals and entities that might prefer to pay on credit or through some mechanism other than cash. Electronic checks are modeled on paper check, accept that they are initiated economically, use digital signatures for signing and endorsing and require the use of digital certificates the authenticate the payer. The payer's bank and bank account.

Electronic checks are delivered either by direct transmission using telephone lines, or by public networks such as the Internet. Electronic check payments (deposits) are gathered by banks and cleared through existing banking channels, such as automated clearing houses (ACH).

8.5.8.1. Benefits of Electronic Checks

Electronic checks have the following advantages :

(*i*) Electronic checks work in the same way as traditional checks, thus simplifying customer education. By retaining the basic characteristics and flexibility of paper checks while enhancing the functionality, electronic checks can be easily understood and readily adopted.

(*ii*) Electronic checks are well suited for clearing micropaymnets; the conventional cryptography of electronic checks makes them easier to process than system based on public-key cryptography (like digital cash). The payee and the payee's and payer's banks can authenticate checks through the use of public-key certificates. Digital signatures can also be validated automatically.

(*iii*) Electronic commerce can serve corporate markets. Firms can use electronic checks to complete payment over the networks in a more cost-effective manner than present alternatives. Further, since the contents of a check can be attached to the trading partner's remittance information, the electronic check will easily integrated with EDI applications, such as account receivable.

(*iv*) Electronic checks create float and the availability of float is an important requirement for commerce. The third party accounting server can earn revenue by charging the buyer or seller a transaction fee or a flat rate fee or it can be act as a bank and provide deposit accounts and make money from the deposit account pool.

(*v*) Electronic check technology links public networks to the financial payment and bank clearing networks.

8.5.1.2. Electronic Checks Action

The electronic check process works in the following way :

Step 1 : Electronic check user must register with the third party account server before they are able to write electronic checks. The account serve also act as a billing service. The registration process can vary depending on the particular account server and may require a credit card or a bank account to back the checks.

Step 2 : Once registered, a consumer can then contact a seller of goods and services.

Step 3 : Using e-mail or other transport methods, the buyer sends an electronic check to the seller for a certain amount of money.

Step 4 : When deposited, a check authorizes the transfer of account balances from the account against which the check was drawn to the account to which the check was deposited.

8.6. DEBIT CARDS BASED EPS

Debit card is the fastest growing payment mode in E-commerce and E-shopping nowadays. Such type of transactions occurs when a customers uses a debit card to make a purchase form the merchant. Customers use these cards at shopping mall, gas station, and other location that accepts them.

The transaction works very similar to credit card system. For example, if we perchage some thing form the shop which accept the debit card then we can make the payment through our debit card, to do so we have to enter our personal pin number given by the some bank. This transacting is routs to bank through ATM network and money transferred form our account to the merchant account.

This transaction taken place within the banks network and safety of the payment is assured. The authentication is provided through some pin or digital network.

8.7. ELECTRONIC FUND TRANSFER

For many consumers, electronic banking means 24-hour access to cash through an automated teller machine (ATM) or Direct Deposit of paychecks into checking or savings accounts. But electronic banking now involves many different types of transactions.

Electronic banking, also known as electronic fund transfer (EFT), uses computer and electronic technology as a substitute for checks and other paper transactions. EFTs are initiated through devices like cards, those we authorize, access our account. Many financial institutions use ATM or debit cards and Personal Identification Numbers (PINs) for this purpose. Some use other forms of debit cards such as those that require, at the most, our signature or a scan.

8.7.1. EFT Services

EFT offers several services that consumers may find practical :

(*i*) Automated Teller Machines or 24-hour Tellers are electronic terminals that let you bank almost any time. To withdraw cash, make deposits, or transfer funds between accounts, we generally insert an ATM card and enter our PIN. Some financial institutions and ATM owners charge a fee, particularly to consumers who don't have accounts with them or on transactions at remote locations.

(*ii*) We may pre-authorize direct withdrawals so that recurring bills, such as insurance premiums, mortgages, and utility bills, are paid automatically.

(*iii*) Pay-by-Phone Systems allows us to call our financial institution with instructions to pay certain bills or to transfer funds between accounts. We must have an agreement with the institution to make such transfers.

(*iv*) Personal Computer Banking allows us to handle many banking transactions via our personal computer. For instance, we may use our computer to view your account balance, request transfers between accounts, and pay bills electronically.

(*v*) Point-of-Sale Transfers allow us to pay for purchases with a debit card, which also may be our ATM card.

(*vi*) Electronic Check Conversion converts a paper check into an electronic payment at the point of sale or elsewhere, such as when a company receives our check in the mail. In a store, when we give our check to a store cashier, the check is processed through an electronic system that captures our banking information and the amount of the check. Once the check is processed, you're asked to sign a receipt authorizing the merchant to present the check to our bank electronically and deposit the funds into the merchant's account.

8.7.2. Suggestions for EFT

If we decide to use EFT, keep these tips in mind :

(*i*) Take care of our ATM or debit card. Know where it is at all times; if we lose it, report it as soon as possible.

(*ii*) Choose a PIN for your ATM or debit card that's different from our address, telephone number, Social Security number, or birthdate. This will make it more difficult for a thief to use our card.

(*iii*) Keep and compare your receipts for all types of EFT transactions with your periodic statements. That way, you can find errors or unauthorized transfers and report them.

(*iv*) Make sure you know and trust a merchant before you share any bank account information or pre-authorize debits to your account. Be aware that some merchants use electronic processing of your check if you sign a receipt authorizing the transaction.

8.8. ELECTRONIC COMMERCE AND BANKING

For many consumers, electronic banking means 24-hour access to cash through an automated teller machine (ATM) or Direct Deposit of paychecks into checking or savings accounts. But electronic banking now involves many different types of transactions.

Electronic banking, also known as electronic fund transfer (EFT), uses computer and electronic technology as a substitute for checks and other paper transactions. EFTs are initiated through devices like cards or codes that allows us to access our account. Many financial institutions use ATM or debit cards and Personal Identification Numbers (PINs) for this purpose. Some use other forms of debit cards such as those that require, at the most, our signature or a scan.

Technology is changing the interaction between bank and customers. In particular, technological innovations have enabled the following capabilities: online delivery of bank brochures and marketing information; electronic access to bank statements; ability to request the transfer of funds between accounts; electronic bill payment and presentation; ability to use multiple financial software products with "memory" (thus eliminating the need to re-enter the same data). Online payments – encrypted credit cards for transferring payment instructions between merchant, bank, customer; and finally micropayments. These online capabilities increase the facility and speed of retail banking.

8.8.1. Changing Dynamics in the Banking Industry

In recent years, there has been a major change in the way banks strive for increased profitability. In the past, the banking industry was chiefly concerned with asset quality and capitalization; if the bank was performing well along these two dimensions, then the bank would likely to be profitable. Today, performing well on asset quality and capitalization is not enough. Banks need to find new ways to increase revenue in a "mature market" for most traditional banking services, particularly consumer credit. A thorough understanding of this competitive environment is needed before banks can determine their online strategy :

- Find distinct factors contribute to the new competitive environment.
- Changing customer needs driven by online commerce.
- Optimization of branch networks in order to reduce costs.
- Changing demographic trends and potential new consumer markets.
- New online financial products.

8.8.2. Changing Consumer Needs

Consumer requirements have changed substantially in the last decade. Customer want to access account-related information, download account data for use with personal finance software products, transfer funds between accounts and pay bills electronically. Of course, along with these services, banks must be able to supply / guarantee the privacy and confidentiality that customer demands, which is not a trivial matter to implement on the part of the banks.

Many consumer requirements are based on the simple premise; customer and financial institutions both seek closer and more multifaceted relationship with one another. Customer want to be able to ban at their convenience, including over the weekend or late at night. Bankers want more stable and long-term relationships with their customers.

8.8.3. Cost Reduction

Since 1984, the banking industry has been consolidating at a rapid pace. The central goal of most mergers is to reduce operating costs. During the decade, 1984-94, the number of banks in the United States fell by 27 percent. A recent survey found that of the 13,000 banks and thrifts operating today, only between 6000 and 9000 institutions will be operating by the year 2000. The reason for this reduction is evident from the economics of existing delivery channels. In general, bricks and mortar branches cost at least $1.5 million each. Online technology can deliver services far more economically than these existing methods, as the infrastructure costs such as PCs are shared with the consumer.

Merill Lynch manages over $500 billion in financial assets and aspects to top $1 trillion by the end of the decade. Merill Lynch also manages more retirement funds - $71 billion worth – than the largest 100 banks combined. Likewise, assets managed by the fidelity Group in the United States have grown $40 billion in 1985 and upto $400 billion in 1994.

8.8.4. Home Banking Implementation Approaches

Pushed growing consumer demand and the fear of losing market shares, banks are investing heavily in home banking technology. Collaboration either hardware, software, telecommunications and other companies, banks are introducing new ways for consumers to access their account balances, transfer funds, pay bills and buy goods and services without using cash, mailing a check or leaving home. The four major categories of home banking (in historical order) are :

(*i*) **Proprietary Bank Dial-up Services :** A home banking service, in combination with a PC and modem (Figure 8.2) lets the bank become an electronic gateway to customer's accounts, enabling them to transfer funds or pay bills directly to creditor's account.

(*ii*) **Off-the-Shelf Home Finance Software** : This category is a key player in cementing relationships between current customers and helping bank gain new customers. Example include Intuit's Quicken, Microsoft Money and Bank of America's MECA software. This software market is attracting interest for bank as it has steady revenue streams by way of upgrades and the sale of related products and services.

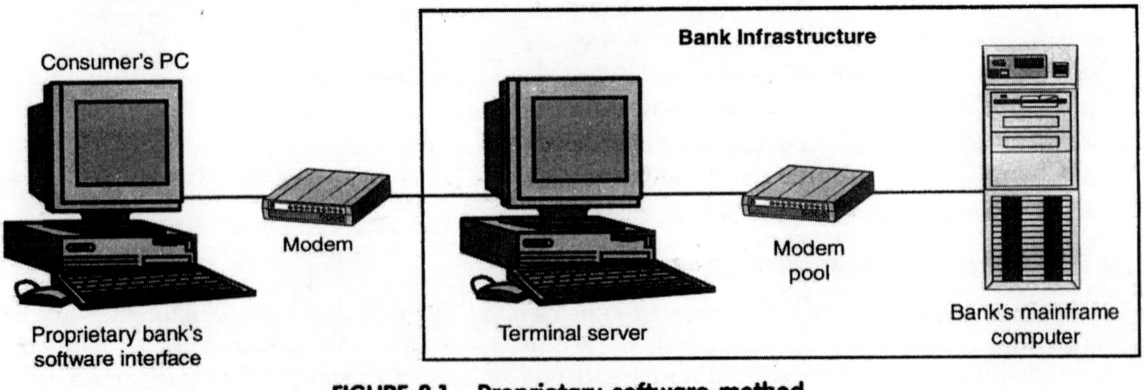

FIGURE 8.1. Proprietary software method.

(*iii*) **Online Serivice-based Banking**: This category allows banks to set up retail branches on subscriber-based online service such as Prodigy, CompuServe and America Online.

(*iv*) **World Wide Web-based Banking**: This category of home banking allows bank to bypass subscriber-based online services and reach the customer's browser directly through the World Wide Web. The advantages of this model are the flexibility at the back-end to adapt to new online transaction processing models facilitated by electronic commerce and the elimination of the constricting intermediary (or online service).

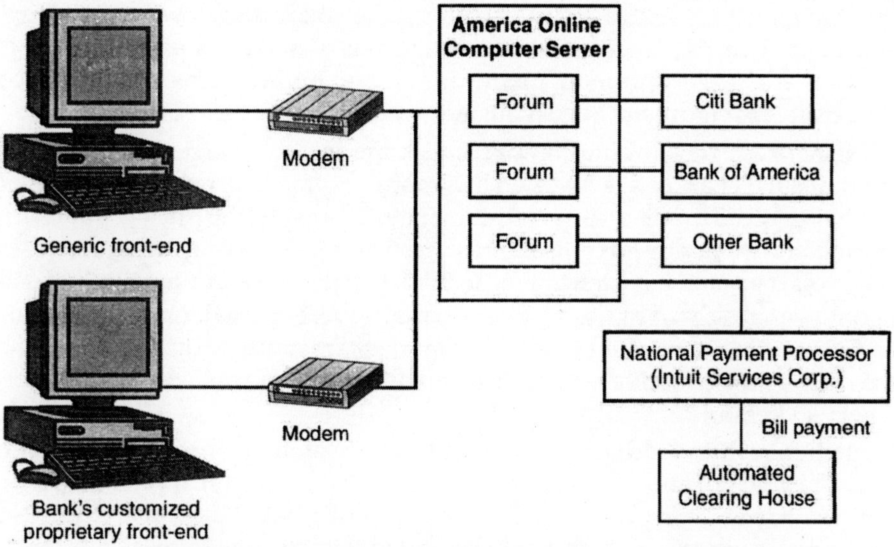

FIGURE 8.2. Online servicing banking architecture.

8.8.5. Management Issues in Online Banking

The challenge facing the Banking Industry is whether management has the creativity and vision to harness the technology and provide customer with new finance products necessary to satisfy their continually changing financial needs. Banks must deliver high quality products at the customers convenience with high-tech, high-touch personal and affordable service. In order to achieve this, management has to balance the five key values that increasingly drive customer's banking decisions: simplicity customized service, convenience, quality and price.

Online banking will realize its full potential when the following key elements fall into place :

(*i*) The development of any interesting portfolio of product and services that are attractive to customers and sufficiently differentiated form competitors.

(*ii*) The creation of online financial supply chains to manage the shift from banks as goalkeeper models to banks as gateways.

(*iii*) The emergence of low-cost interactive access terminals for the home as well as affordable interactive home information services.

(*iv*) The identification of new market segments with untapped needs such as the willingness to pay for the convenience of remote banking.

(*v*) The establishment of good consumer service on the part of banks. The fact that technology increases the ease of switching from one bank to another means that banks that do not offer superior customer service may see low level of customer loyalty.

(*vi*) The development of effective back-office systems that can support sophisticated retail interfaces.

8.9. ELECTRONIC COMMERCE AND ONLINE PUBLISHING

Initially, growth in the online publishing marketplace was driven by the potential of the new interactive technologies and applications. The promise of new interactive publishing captured the imagination of both content providers and the public. However, from 1993 to 1995 much of online publishing was inhibited by a lack of business purpose.

Many online publishing pioneers have gone up to the technology curve and are confronting tough management questions such as how to gain market share and how to be profitable sooner than later. Some of these firms has invested tens of millions of dollars in payroll alone. Many early pioneers have invested a large amount of money into brand building, marketing and content, but they have not been able to figure out which business model works best for making money.

Online publishers are developing new business models to charge customers directly and convince them that such charges are justified. As more and more firms are begin to offer online content, they are being forced to adjust to new customer attitude regarding pricing. Publishers currently finance their businesses by offering advertisers mass markets for delivering their message in return for large advertising fees. The public has been trained to think that the news, information and entertainment they receive should be subsidized or nearly free and that advertisers will pay the bill. This approach may not be viable in the online medium when mass markets are replaced by customers selecting their information and delivery methods. For instance, early experiments with online-subscriber based services have not been very successful.

Today, publisher such as Addison Wesley only offer catalogs or sample selections of work available online. They do not and cannot offer more because in an environment where the culture and technology provide so little protection for the rights of content producer, there is too great a risk to their intellectual property. The Internet makes it extremely easy to copy, retransmit and alter works without the permission of the copyright holder. Moreover, the digital world has no international boundaries and policing is impossible since the levels of protections and sanctions against infringement vary widely in countries across the globe, which makes the risk even greater. Clearly without effective production, publishers are not willing to risk their investment and hard work.

8.9.1. Need of Online Publishing

The recently the emergence of the Internet and Web have played a large role in bringing online publishing from the niche target audience to a mass audience. Some of the reasons for the dramatic and sudden increase in the Internet of publishers, advertisers and content providers in the Web include :

Widespread adoption of the web combined with rapidly evolving technology, makes it possible to meet time-to-market requirements, that were unimaginable even a few years ago.

Much of the investment in infrastructure has been made by others: Internet access providers, hardware companies and various software developers. Huge investments in advance of uncertain returns are not necessary.

The web allows new ways of targeting and reaching customers. This also coincides with publisher's desires to migrate from talking to large numbers of customer in a mass market to working with smaller target groups.

The publishing community's rapid movement toward the web seems to have caught the advertising world by surprise. Advertising is an industry that has been relatively untouched by the technology.

Needless to say, the shift in the publishing world has brought forward a massive flow of commentary, forecasts and philosophical and sociological musings as well as road maps of the future for investors and other interested parties. These debates can be categorized into three main areas. On one side are technical issues :

(*i*) How to improve and enforce copyright protection ?

(*ii*) How to create compelling content using the available technology ?

(*iii*) How to conduct financial transaction for individual articles (micropayments) ?

On the other side are management issues :

(*i*) How to create an organization for online publishing ?

(*ii*) How to select business models that ensure success in different segments of online publishing ?

(*iii*) How to satisfy online customers and keep them loyal ?

(*iv*) In the middle is the more philosophical debate on what exactly is meant but online publishing.

Online publishing can be viewed in two ways :

(*a*) As the activity of disseminating information (which scholars have done for thousand of years and purpose for which the web was invented).

(*b*) As the activity of publishing for sale. In order to realize revenue from their information, publishers package their product as a bundled commodity rather than disjointed information.

This is a fundamental difference between information and commodities. A commodity such as digital book need to be sold in volume or order to generate revenue, so publishers must keep their material from spreading freely all over the planet. On the other hand, the value of information is not lost when it is sold because the contents are not limited by their medium of expression.

8.9.2. Online Publishing Approach

There are four contrasting content publishing approach :

1. **The Online Archive Approach** : This is new to the web but is a logical extension of the trends in electronic delivery over the past several years.
2. **The New Medium Approach** : This is more controversial and more difficult to implement but also more exciting.
3. **The Publishing Intermediation Approach** : This is an online extension of the third-party publisher role-off line.
4. **The Dynamic and Just-in-Time Approach** : In this approach content is assembled in real-time and transmitted in the format best suited to the user's tastes and preferences.

8.10. ELECTRONIC COMMERCE AND RETAILING

Almost every retailer is reevaluating every aspect of its operation from customer service to advertising, merchandising to store design and logistics to order fulfillment. Furthermore, reaching to the pressure of retailers, suppliers and assessing technology-based solutions to drive down costs (labour, delivery and production) and become more efficient producers of goods.

Online channels such as online services and the web are also impacting traditional retail business models. In the traditional model, the customer went to the store and located the product. In the online model, the retailer seeks out the customer. The success of catalog retailers demonstrates that a significant portion of customers have embraced the reverse model: the retail going to the customer.

However, retailers need to consider the following issues in developing a business model :

(i) **Product/Content Issues :** What kind of product are suited for online retailing ?

(ii) **Software Interface Issues :** What kind of feature will constitute an effective interface ? What features make it easy to find and select items for online purchase ?

(iii) **Process Issue :** What are the specific steps in the shopping process from a consumer's perspective? What kind of processes should companies develop to fulfill orders efficiently ?

(iv) **Pricing Issues :** How much will consumers be willing to pay for convenience ?

(v) **Payment Issues :** What payment methods would consumer use for online purchases ?

(vi) **Market Penetration Issues :** would an online channel be popular with consumers ? How long will such a system take to attract a critical mass and become profitable ?

Understanding these issues will help companies formulate a merchandising strategy – and build a merchandising mindset – from what they sell to how they sell it; from pricing to timing; from product to development to sourcing; from their choices of electronic media to how they integrate them; and from customer payment alternatives to shipping procedures. We need critical analysis of these questions in order to maintain the proper perspective on online retailing.

8.11. ONLINE SHOPPING

Online retailers can be divided into four distinctive groups :

(a) Retailers that operate offline stores, getting their feet wet with online storefronts.

(b) Retailers such as catalogers who are concerned with the new web-based methods of catalog retailing.

(c) "New age" retailers that do not currently operate retail businesses, but they understand how this new medium can revolutionized shopping.

(d) Manufacturers exploring the potential to go "direct to consumers".

Retailing via online services has been around for a while. Online service providers such as CompuServe and American Online (AOL), offers twenty four hours access to their own shopping environments. These online malls can be very effective. For instance, when home shopping Network (HSN) launched its retail outlet on Prodigy in late 1994, subscriber purchased the entire inventory of cookware and jewelry in only a few days. Another example can be found in the music business, tower records. Tower jumped online in June 1995, when it put a store up on AOL. In December 1995, the tower's AOL store registered almost $200,000 in sales, exceeding tower's expectations in terms of sales.

The intriguing aspect of online retailing is the emergence of companies that only exist online like the Electronic Newstand, Amazon.com, Pawws and E-Trade. The brief history of E-Trade securities illustrates what the new paradigm is all about. E-Trade is advantages that few Internet Companies can claim: profits and operating history. E-Trade started accepting stock trades on a web page. The impact on the business could be seen within two months: about 10 percent of its trade came through the web. The ability to be online made pricing more competitive. E-Trade charges $14.95 for a listed market order. The success of this firm indicates that there is a significant market potential online waiting to be tapped into.

SUMMARY

In this chapter, we have gone through various mode of electronic payment system like credit cards, debit cards and EFT. SET (Secure Electronic Transaction) standard was developed for ensuring the transaction security of information on internet. Nowadays banking industry has change through e-banking, E-commerce is giving a new shop to online publishing, Retailing and online shopping.

REVIEW QUESTIONS

1. How electronic payment systems differs with traditional payment systems.
2. What is SET and how SET can be implemented to make our transaction more secure ?
3. What do you understand by payment gateways and how payment gateways work ?
4. How online credit card based system are used in online transactions ?
5. Write the short notes on following :
   (*i*) Virtual PIN
   (*ii*) Digital Cash
   (*iii*) Cyber Cash
   (*iv*) Paypal
   (*v*) Smart Cards
   (*vi*) Debit Cards.
6. How electronic checks are advantageous over the traditional checks ?
7. How online banking is giving a push for e-commerce ?
8. What is the need of online publishing ?
9. How retail industry is utilizing the e-commerce ?
10. "Now a days on-line shopping". Changing the market dynamics, comment on it ?

❑❑❑

9

CHAPTER

NET COMMERCE

INSIDE THIS CHAPTER

9.1. Electronic Data Interchange
9.2. Supply Chain Management
9.3. Customer Relations Management (CRM)

9.1. ELECTRONIC DATA INTERCHANGE

EDI developed in the 1960s as a mean of accelerating the movement of documents pertaining to shipments and transportation. Not until the mid-1980s, however, was the technique used in a wide range of industries – automotive, retail, transportation and international trade. Its use is growing and it is set to become the standard by which organization will communicate formally with each other in the word of electronic commerce.

Electronic commerce is often equated with EDI, so it is important to qualify that electronic commerce adopts EDI and much more. In electronic commerce, EDI techniques are aimed at improving the interchange of information between trading partners, suppliers and customers.

Defining EDI

Because of the different approaches in the development and implementation of EDI, there is no consensus on a definition of EDI. A review of some of the prevailing definitions follows:

"Electronic Data Interchange is the transmission, in a standard syntax, of unambiguous information of business or strategic significance of computers of independent organizations. [*The Accredited Standard Committee for EDI of the American National Standards Institute*]."

"Electronic Data Interchange is the interchange of standard formatted data between computer application systems of trading partners with minimal manual invention. [UN/EDIFACT *Training Guide*]".

"Electronic Data Interchange is the electronic transfer, from computer to computer of commercial and administrative data using an agreed standard to structure an EDI message."

"Electronic Data Interchange is the electronic transfer from one computer to another of computer processable data using an agreed standard to structure the data."

Another aspect of EDI that often causes confusion is its usage in one context as a technological solution that focuses on the mechanical transport and assembly of business forms and in other context as a business methodology that focuses on the content and structure of forms. This confusion can be cleared by examining the layered architecture of EDI.

9.1.1. EDI Layered Architecture

EDI has semantic Application Layer, Translation Layer, Packaging and Infrastructure Layer.

The EDI semantic layer describes the business application that is driving EDI. For a procurement application, this translates into requests for quotes, price quotes, purchase orders, acknowledgement and invoices. This layer is specific to a company and the software it uses. In other words, the user interface and content visible on the screen arc tailored and customized to local environments.

The information seen at the EDI semantic layer must be translated from a company-specific form to a more generic or universal form so that it can be sent to various trading partners, who could be using a variety of software applications at their end.

To facilitate the transfer of computer files between two "trading partners" requires that the computer application of both sender and receiver use a compatible format for EDI document exchange. The sender must use a software application that creates an EDI file format similar to what the recipient's computer application can read. It is not mandatory that both have identical file processing systems. When the trading partner sends a document, the EDI translation software converts the proprietary format into a standard mutually agreed on by the processing systems. When a company receives the document, their EDI translation software automatically changes the standard format into the proprietary format on their document processing software so that the company can manipulate the information in whatever way it chooses to.

EDI *standards* specify business form structure and to some extent influence content seen at the application layer. For instance, a purchase order name field in an X.12 standard might be specified to hold a maximum of 50 characters. An application using 75-character field length will produce name truncation during the translation from the application layer to the standard layer. In short, the EDI standards and application level, although separate, are closely intertwined.

The **EDI transport** *layer* corresponds closely with the nonelectronic activity of sending business form from one company A to company B. The business form could be sent via regular postal service, registered mail, certified mail, to private carrier or simply faxed between the companies.

EDI document transport is far more complex than simply sending e-mail messages or sharing files through a network, a modem or a bulletin board. These EDI documents are more structured than e-mail and typically are manipulated or processed more than e-mail messages by the sending and receiving software.

9.1.2. EDI Versus E-mail

Electronic Data Interchange (EDI)	Electronic Mail
There is typical no human involvement in the processing of the information, as the interface has software-to-software orientation.	The data are structured in a software-understandable way.
The interchange is composed by one software for interpretation by another software. If a reply is involved, it is composed by a software to be interpreted by another software.	The data are not necessarily structured to be software-understandable. A human-to-software interface is involved at a minimum of one end of the interchange.

The message is composed by a human and/or interpreted by a human and/or a reply is composed by a human and/or interpreted by a human.

9.1.3. EDI in Action

The idea behind EDI is very simple. EDI seeks to take what has been a manually prepared form or a form from business application, translate that data into a standard electronic format and transmit it. At the receiving end, the standard format is "untranslated" into a format that can be read by the recipient's application. Hence output of one application become input to another through the computer-to-computer exchange of information. The result is an elimination of the delays and the errors inherent in paper-based transactions.

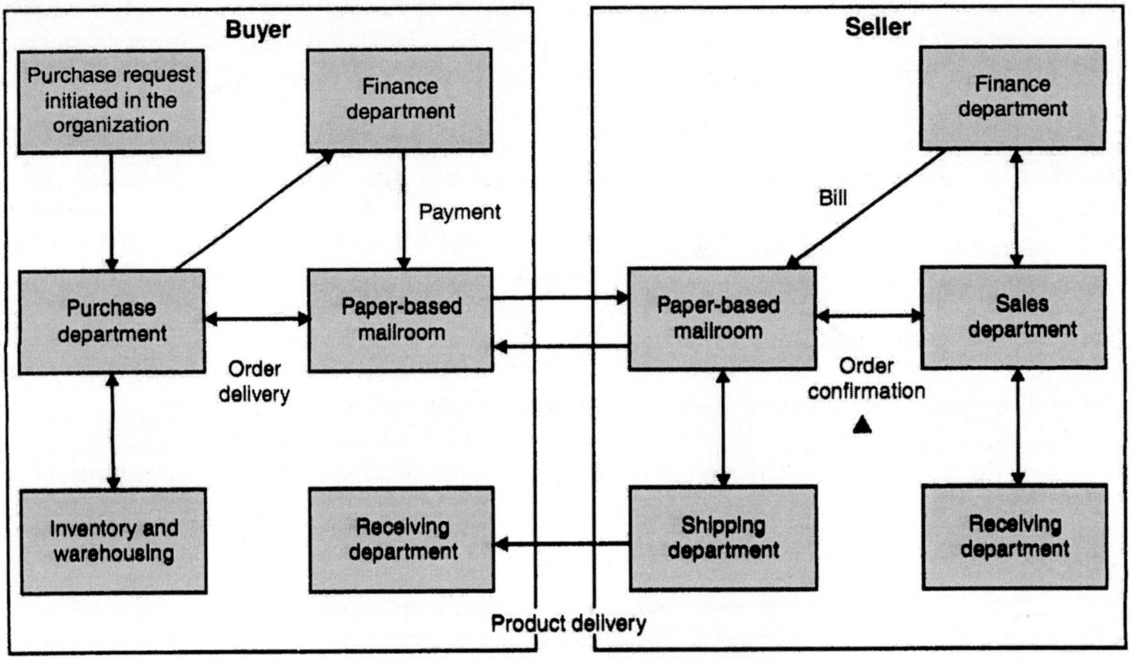

FIGURE 9.1. Information flow through EDI.

Benefits of EDI can be seen by comparing the flow of information between organizations before and after its implementation. For the purpose the purchasing application provides an ideal scenario. In general, EDI has been used extensively in the procurement function to streamline the interaction between the buyer and seller. Other uses for EDI to exchange transcript quickly. Auto manufacturers use EDI to transmit large, complex engineering designs created on specialized computers. Large multinational firms use EDI to send online price catalogs to customers listing products, prices and terms and make direct orders by EDI.

Figure 9.1 shows the information flow when the paper document are shuffled between organizations via the mailroom.

EDI can substantially automate the information flow and facilitate management of the business process, as illustrated in Figure 9.1. The EDI transactions for a purchase, shipment and corresponding payment are as follows:

Step 1. Buyer's computer sends *purchase order* to seller's computer.

Step 2. Seller's computer sends *purchase order confirmation* to buyer's computer.

Step 3. Seller's computer sends *booking request* to transport company's computer.

Step 4. Transport company's computer sends *booking confirmation* to seller's computer.

Step 5. Seller's computer sends *advance ship notice* to buyer's computer.

Step 6. Transport company's computer send *status* to seller's computer.

Step 7. Buyer's computer sends *Receipt advice* to seller's computer.

Step 8. Seller's computer sends *invoice* to buyer's computer.

Step 9. Buyer's computer sends *payment* to seller's computer

The purchase order confirmation is the seller's acceptance of the price and terms of sale. Note that the various internal departments are agreed and called buyer and seller to simplify the description. All the interaction occurs through EDI forms and in most cases are generated automatically by the computer.

Tangible Benefit to EDI

Savings can be achieved from the following improvement :

(*i*) **Reduced Paper-Based System :** EDI can impact the effort and expense a company devotes to maintain records, paper-related supplies, filing cabinets or other storage systems and to the other personnel required to maintain all of these systems. Electronic transactions take over most of the functions of paper forms and through automation drastically reduce the time spent to process them.

(*ii*) **Improved Problem Resolution and Customer Service :** EDI can minimize the time companies spend to identify and resolve interbusiness problems. Many such problems come from data entry errors somewhere along the way and EDI can eliminate many of them.

(*iii*) **Expanded Customer/Supplier Base :** Many large manufacturers and retailers with the necessary clout are ordering their suppliers to institute an EDI program. However, these are isolated island of productivity because they are unable to build bridges to other companies. With the advent of electronic commerce, the bridge is now available.

9.1.4. Financial EDI

Financial EDI comprises the electronic transmission of payment and remittance information between a payer and their respective banks. Here we will examines the way business-to-business payments are made today and describes the various methods for making financial EDI payments.

Financial EDI allows businesses to replace the labour-intensive activities associated with issuing, mailing and collecting checks through the banking system with automation initiation, transmission and processing of payment instructions. Thus it eliminates the delays inherent in processing checks. Financial EDI also improves the certainty of the payment flows between corporate bank accounts because the payee's bank can credit its account on the scheduled payment date and the payer's can debit its account on the same day.

9.1.4.1. Types of Financial EDI

Traditionally, wholesale or business-to-business payment is accomplished using checks, EFT and automated clearinghouse (ACH) for domestic and international fund transfer. ACH provide two basic services to industrial and financial corporate customers (including other banks) :

(1) Fast transmission of information about their financial balances throughout the world and

(2) Movement of money internationally by rapid speed for settlement of debit/credit balances. Banks have developed sophisticated cash management systems on the back of these services that essentially reduce the amount of money companies leave idly floating in low-earning accounts.

Let's provide a quick overview of the three principal types of noncash payment instruments currently used for business-to-business payments; **checks, electronic fund transfer and automated clearinghouse (ACH) transfers.**

1. Bank Checks

Checks are instruments for debit transfer where payees collect fund from payers. Funds made available by banks to depositors of checks are provisional and may be reversed if the payer does not have sufficient funds in its account to pay the check when it is received by the payer's bank.

2. Interbank Electronic Funds Transfer (EFT)

Electronic Funds Transfer (EFT) are credit transfer between banks where fund flow directly from the payer's bank to the payee's bank. They are same day almost instantaneous payments. EFT is one of the earliest example of payment systems that use online transactions, although these transactions are carried out on private networks.

3. Automated Clearing House (ACH) Transfers

In contrast of the EFT process, ACH transfers are used to process high volumes of relatively small-dollar payments for settlement in one or two business days. An ACH provides the following services: preauthorized credits such as the direct deposit of payrolls; preauthorized debits such as repetitive bill payments; and consumer-initiate payments (called GIRO in baking circles).

9.1.5. Standardization and EDI

Connecting trading partners is not enough for EDI. All software, hardware and networks must work together so that information flows from one source to another. Indeed, the computer industry has spawned a multitude of de facto, often overlapping standards, because many independent groups develop them. The result has been slow deployment and restricted implementation of EDI between companies.

The advantage of standardized is numerous. One field that has benefited immensely from standardization is the electronics industry. In the 1960s and 1970s electronic components were designed without concern for compatibility. This created a lot of confusion and spurred the institute of electrical and electronic engineers (IEEE) to issue guideline standards for capacitors, resistors, power supplies and all other components. As components become substitutable, production boomed, competition increased and price deflated. Likewise, standardization is expected to promote "interoperability" between EDI implementations.

Today, two major EDI standard exist: the American National Standard Institute (ANSI) X.12 Committee and the United Nations EDI for Administration, Commerce and Trade (EDIFACT) standards fro-international usage.

9.1.5.1. EDIFACT

Developed by the United Nations, EDIFACT is a family of standards similar to ANSI X.12. EDIFACT was based on TRADECOMS, developed by the U.K. department of Customs and Excise with the assistance of SITPRO (The British Simplification of Trade Procedures Board).

It is further developed by the United Nations Economic Commission for Europe – Working Party (four) on facilitation of International Trade Procedures (UN/ECE/WP4). EDIFACT is widely accepted as the foremost international EDI standard. Today, EDIFACT and ANSI are working towards compatibility.

9.2. SUPPLY CHAIN MANAGEMENT

Supply chain management (SCM) is the oversight of materials, information, and finances as they move in a process from supplier to manufacturer to wholesaler to retailer to consumer.

Supply chain management involves coordinating and integrating these flows both within and among companies. It is said that the ultimate goal of any effective supply chain management system is to reduce inventory (with the assumption that products are available when needed). As a solution for successful supply chain management, sophisticated software systems with web interfaces are competing with web-based application service providers who promise to provide part or all of the SCM service for companies who rent their service.

Supply chain management flows can be divided into three main flows:

1. The product flow
2. The information flow
3. The finances flow

The product flow includes the movement of goods from a supplier to a customer, as well as any customer returns or service needs. **The information flow** involves transmitting orders and updating the status of delivery. **The financial flow** consists of credit terms, payment schedules, and consignment and title ownership arrangements.

9.2.1. Basic Component of SCM

Supply chain management (SCM) is the combination of art and science that goes into improving the way your company finds the raw components it needs to make a product or service and deliver it to customers. The following are five basic components of SCM.

1. Plan : This is the strategic portion of SCM. We need a strategy for managing all the resources that go toward meeting customer demand for our product or service. A big piece of planning is developing a set of metrics to monitor the supply chain so that it is efficient, costs less and delivers high quality and value to customers.

2. Source : Choose the suppliers that will deliver the goods and services we need to create our product. Develop a set of pricing, delivery and payment processes with suppliers and create metrics for monitoring and improving the relationships. And put together processes for managing the inventory of goods and services you receive from suppliers, including receiving shipments, verifying them, transferring them to your manufacturing facilities and authorizing supplier payments.

3. Make : This is the manufacturing step. Schedule the activities necessary for production, testing, packaging and preparation for delivery. As the most metric-intensive portion of the supply chain, measure quality levels, production output and worker productivity.

4. Deliver : This is the part that many insiders refer to as logistics. Coordinate the receipt of orders from customers, develop a network of warehouses, pick carriers to get products to customers and set up an invoicing system to receive payments.

5. Return : The problem part of the supply chain. Create a network for receiving defective and excess products back from customers and supporting customers who have problems with delivered products.

9.2.2. Supply Chain Management Software

Supply chain management software is possibly the most fractured group of software applications on the planet. Each of the five major supply chain steps previously outlined composes dozens of specific tasks, many of which have their own specific software. Some vendors have assembled many of these different chunks of software together under a single roof, but no one has a complete package that is right for every company. For example, most companies need to track demand, supply, manufacturing status, logistics (i.e. where things are in the supply chain), and distribution. They also need to share data with supply chain partners at an ever increasing rate. While products from large ERP vendors like SAP's Advanced Planner and Optimizer (APO) can perform many or all of these tasks, because each industry's supply chain has a unique set of challenges, many companies decide to go with targeted best of breed products instead, even if some integration is an inevitable consequence.

It's worth mentioning that the old adage about systems only being as good as the information that they contain applies doubly to SCM. If the information entered into a demand forecasting application is not accurate then we will get an inaccurate forecast. Similarly, if employees bypass the supply chain systems and try to manage things manually, then even the most expensive systems will provide an incomplete picture of what is happening in a company's supply chain.

9.2.2.1. Roadblocks to Installing Supply Chain Software

Let us discuss some of roadblocks in SCM software:

1. **Gaining Trust from Our Suppliers and Partners :** Supply chain automation is uniquely difficult because its complexity extends beyond our company's walls. Our people will need to change the way they work and so will the people from each supplier that you add to your network. Only the largest and most powerful manufacturers can force such radical changes down suppliers' throats. Most companies have to sell outsiders on the system. Moreover, our goals in installing the system may be threatening to those suppliers, to say the least. For example, Wal-Mart's collaboration with P&G meant that P&G would assume more responsibility for inventory management, something retailers have traditionally done on their own. Wal-Mart had the clout to demand this from P&G, but it also gave P&G something in return-better information about Wal-Mart's product demand, which helped P&G manufacture its products more efficiently. To get our supply chain partners to agree to collaborate with us, we have to be willing to compromise and help them achieve their own goals.

2. **Internal Resistance to Change :** If selling supply chain systems is difficult on the outside, it isn't much easier inside. Operations people are accustomed to dealing with phone calls, faxes and hunches scrawled on paper, and will most likely want to keep it that way. If we can't convince people that using the software will be worth their time, they will easily find ways to work around it. We cannot disconnect the telephones and fax machines just because we have supply chain software in place.

3. **Many Mistakes at First :** New supply chain systems process data as they are programmed to do, but the technology cannot absorb a company's history and processes in the first few months after an implementation. Forecasters and planners need to understand that the first bits of information they get from a system might need some tweaking. If they are not warned about the system's initial naiveté, they will think it is useless. In one case, just before a large automotive industry supplier installed a new supply chain forecasting application to predict demand for a product, an automaker put in an order for an unusually large number of units. The system responded by predicting huge demand for the product based largely on one unusual order. Blindly following the system's numbers could have led to inaccurate orders for materials being sent to suppliers within the chain. The company caught the problem but only after a demand forecaster threw out the system's numbers and used his own. That created another problem: Forecasters stopped trusting the system and worked strictly with their own data. The supplier had to fine-tune the system itself, then work on reestablishing employees' confidence. Once employees understood that they would

be merging their expertise with the system's increasing accuracy, they began to accept and use the new technology.

9.2.3. Supply Chain Collaboration

Let's look at consumer packaged goods for an example of collaboration. If there are two companies that have made supply chain a household word, they are Wal-Mart and Procter & Gamble. Before these two companies started collaborating back in the '80s, retailers shared very little information with manufacturers. But then the two giants built a software system that hooked P&G up to Wal-Mart's distribution centers. When P&G's products run low at the distribution centers, the system sends an automatic alert to P&G to ship more products. In some cases, the system goes all the way to the individual Wal-Mart store. It lets P&G monitor the shelves through real-time satellite link-ups that send messages to the factory whenever a P&G item swoops past a scanner at the register.

With this kind of minute-to-minute information, P&G knows when to make, ship and display more products at the Wal-Mart stores. No need to keep products piled up in warehouses awaiting Wal-Mart's call. Invoicing and payments happen automatically too. The system saves P&G so much in time, reduced inventory and lower order-processing costs that it can afford to give Wal-Mart "low, everyday prices" without putting itself out of business.

9.2.4. Extended Supply Chain

The extended supply chain is a clever way of describing everyone who contributes to a product. So if we make text books, then our extended supply chain would include the factories where the books are printed and bound, but also the company that sells us the paper, the mill where that supplier buys their stock, and so on. It is important to keep track of what is happening in your extended supply chain because with a supplier or a supplier's supplier could end up having an impact on us (as the old saying goes, a chain is only a strong as its weakest link). For example, a fire in a paper mill might cause the text book manufacturer's paper supplier to run out of inventory. If the text book company knows what is happening in its extended supply chain it can find another paper vendor.

9.2.5. Impact of Globalization on the Supply Chain

Just in time manufacturing isn't the only way companies have used their supply chains to reduce cost. Manufacturing in developing countries is substantially cheaper than in the U.S. because of the low cost of labor. For example, the median wage at a Chinese manufacturing plant is 1,000 yuan, or about $120, per month, according to a 2005 survey by The MPI Group. But foreign manufacturing brings with it another set of challenges. It isn't as easy to set up real-time data sharing with a factory in, say, China as it is with a factory we own in the United States. And the sheer distance that overseas goods need to travel—not to mention the number of vessels they need to travel on – in order to reach the U.S. increases the chance that they will get delayed. The bottom line is that foreign manufacturing brings back a lot of the uncertainty that supply chain systems were designed to eliminate. The good news is that technology capable of tracking shipments throughout the world is getting better. The bad news is that a lot of this technology is still pretty expensive, that some of the places you would want to deploy it don't have the necessary infrastructure in place, and well, there isn't a piece of technology out there that can make up for the whim of a Chinese Customs official. Furthermore, labor costs in some places are so low that IT automation and monitoring projects may add more to costs—in terms of software, hardware and still-precious (and unreliable) bandwidth—than they save in productivity. Hence, some low-tech or commodity products may not be worth monitoring at all until they hit a ship in a foreign port.

In the meantime, the best bet is to use whatever systems we can to gain as much visibility into the global supply chain as possible. It may be impossible to replicate the just in time model on a global scale, but by applying whatever technology we can, and by choosing the supply chain partners who have the capability to share data with us, we can get many of the benefits of just in time while paying low foreign prices.

9.2.6. Some Emerging Technologies that will Affect the Supply Chain

The most notable is Radio Frequency Identification, or RFID. RFID tags are essentially barcodes on steroids. Whereas barcodes only identify the product, RFID tags can tell what the product is, where it has been, when it expires, whatever information someone wishes to program it with. RFID technology is going to generate mountains of data about the location of pallets, cases, cartons, totes and individual products in the supply chain. It's going to produce oceans of information about when and where merchandise is manufactured, picked, packed and shipped. It's going to create rivers of numbers telling retailers about the expiration dates of their perishable items—numbers that will have to be stored, transmitted in real-time and shared with warehouse management, inventory management, financial and other enterprise systems. In other words, it is going to have a really big impact.

Another benefit of RFIDs is that, unlike barcodes, RFID tags can be read automatically by electronic readers. Imagine a truck carrying a container full of widgets entering a shipping terminal in China. If the container is equipped with an RFID tag, and the terminal has an RFID sensor network, that container's whereabouts can be automatically sent to Widget Co. without the truck ever slowing down. It has the potential to add a substantial amount of visibility into the extended supply chain.

Right now the two biggest hurdles to widespread RFID adoption are the cost of building the infrastructure and the lack of agreed-upon industry standards.

9.3. CUSTOMER RELATIONS MANAGEMENT (CRM)

CRM is a much misused word since it was hijacked by IT companies to describe IT systems for sales, customer care, and marketing promotions. Clearly, this is only one aspect of CRM

The larger and more holistic view of CRM, 'managing your customer,' involves a cohesive view of branding, promotions, communication, customer touch points—and may very well include issues such as packaging. In many companies, there is still confusion about what CRM is all about. To some, it is about a loyalty scheme or about a help desk. To others, it is about a relational database for key account management or about mass profiling the customer base without undertaking detailed segmentation. Relatively, few organizations have implemented an integrated approach which addresses all the key strategic elements of CRM.

Basically CRM means understanding customers using quantitative and qualitative research, segmenting them and segregation of each of the segments based on their expectations and contribution to profits.

9.3.1 People Aspect of CRM

What should be communicated to the front-line employees to ensure that the 'segment-promise' is indeed communicated in each and every transaction that occurs with customers ? How should they be trained ? What should be the organization structure to allow for quick decision-making on the frontline ? How should these employees be evaluated and rewarded ? How should the back-office support be evaluated and rewarded so that they provide the consistent level of support that is expected for each segment ?

9.3.2. Process and Technology Aspects to CRM

The delivery of all touch-points needs to be envisioned with the 'segment-promise' in mind. Typically, customers look for consistency of delivery. That is where tools like six-sigma help in not just defining processes, but measuring them to assess consistency in delivery. While simple transactions would have a base-level performance, more complex transactions would be tailored per segment.

The problem with implementing technology in these spheres is that features sometimes overwhelm the overall objective. Therefore, technology needs to be envisioned along with processes and they should be seamlessly implemented to ensure base level and segment level promises are implemented consistently.

One of the key challenges that organizations need to overcome is the product-centricity that is inherent within them. Organizations need to become customer-centric. This will allow for consistent interactions with their most important constituency—customers.

9.3.3. Issues of CRM

When the time comes to select a CRM software solution, the project team should create a long list with all eligible vendors, boil it down to a short list, and formally evaluate the short-listed vendors. A formal evaluation should evolve around key questions related to CRM software:

1. **Support of Core Processes :** What is the ability of the vendor to support the core processes in my industry or my company ? Does the vendor provide industry-specific functionality and pre-configurations, best practices, or templates and implementation guidelines ? Is the solution able to handle special issues related to how I sell, market and service, e.g. issues related to resellers, dealers, partners, or agents, or complex product configuration and pricing issues, etc. ?
2. **Total Cost of Ownership :** What is the estimated total cost of ownership of the CRM solution over the expected useful life of the solution ? How many internal IT and business resources do we need to budget with for implementation and ongoing support ? How fast will I be able to show a quick win to my manager and project stakeholders ?
3. **Ease of Use :** How does key user groups, such as sales, marketing, service or support staff respond to the solution ? Can they see obvious benefits from using the application or do they regard it as cumbersome and unproductive ? What possibilities does the solution offer in terms of increasing user adoption rate ?
4. **Ease of Knowledge Transfer :** How complex is it for my internal staff to assume ownership of the solution ? When will they be able to execute configuration changes, create additional functionality or process flows, or rollout the solution to new units or user groups ? Are the components based on proprietary, non-common technologies, or they based on commonly available technologies and open standards ?
5. **Ease of Integration :** Does the CRM software support automated workflows and processes involving non-CRM applications, such as applications related to order management, billing, accounts receivable, inventory, and service management ? How difficult is any required integration to other applications from a technical and semantic point of view ?
6. **Support of Business Intelligence :** What does the solution offer in terms of pre-packaged analytics and reporting ? Can the relevant managers get the information they need in time and good quality ? How easy is it for local managers to develop needed reports and alerts ?
7. **Availability of External Resources :** What is the availability of support staff, consultants, resellers, and references in my local region or country ? What global resources, such as

user groups, web-based forums, and phone-based support are available should I become a customer ?

8. **Level of Security :** How secure is the confidential customer information both from external threats and internal threats, such as disgruntled employees. How well does the solution support differentiated user profiles to ensure that employees and external users only access what they need and are allowed to access ?

These eight points are generally applicable across most CRM software selections. However, criteria might change as one project might have significant focus on mobile users in which case support of mobile users becomes a key issue to consider.

SUMMARY

Basically now-a-days e-commerce is a big hit and every wants to use it, so lot of social and legal issues arises in it. These issues must be addressed for the successful implementation of e-commerce. Nowadays retail distribution is managed through supply chain management. Organization and corporates are using CRM software to increase interaction with their customers.

REVIEW QUESTIONS

1. What is EDI and how it is different from E-mail ?
2. Discuss the Layered architecture of EDI.
3. What is the role of ethics in E-commerce ?
4. "There is a threat on privacy in e-commerce". How we can overcome to this problem ?
5. What are the various legal options in e-commerce for it's implementation ?
6. "United Nations plays an important role to make standard for e-commerce". Justify it ?
7. What are various legal options are available in India about e-commerce ?
8. What do you understand by SCM and how it help in business ?
9. What do you understand by CRM and how it helps in business ?
10. What are various management issues in CRM ?

❑❑❑

CHAPTER 10

E-COMMERCE ISSUES, LAWS AND POLICIES

10.1. SOCIAL ISSUES IN E-COMMERCE

INSIDE THIS CHAPTER

10.1. Social Issues in E-Commerce
10.2. Privacy Issue in E-Commerce
10.3. E-Commerce Laws
10.4. E-Commerce: Government's Policies and Agenda

The question ethics in e-commerce is the current challenge confronting various organizations. IT professionals and IT managers are those in disciplines like medicine and law subscribe to code of ethics that govern the way they behave with clients, customers and the society at large. Trust is linked to the expectation that a professional will behave ethically. This is essential in business because our society is depend on fairness and good judgment. Business people are expected to tell the truth and warn costumers when a fault is discovered in a product.

10.1.1. Definition of Ethics in E-Commerce

Definition of Ethicks in E-commerce are as follows : Ethics is one or all of the following: *fairness, justice, equity, honesty, trustworthiness* and *equality*. Stealing, cheating, lying or backing out on one's word all describe a lack of ethics. Something is ethical when a person feels it is innately right, which is a subjective judgment.

Figure 10.1 is a conceptual model of a acceptable behaviour, with ethics as a factor in each quadrant. For example, if a person falsely reports a donation to a Charitable Organization, it is considered both illegal and immortal (quadrant A). This action implies a lack of ethical standards and personal values. In quadrant D, rescuing hostages from a foreign country might be illegal (and dangerous) by the country's law, but it is moral (and implicitly ethical) under laws. Deciding what is moral, legal or ethical is a constant struggle.

An unethical act is not the same as an immoral or an illegal act, although one may lead to the other. For example :

1. Cheating on a Income Tax return is more illegal than immoral, although it is implicitly unethical.
2. Cheating on a friend is more immoral that illegal, although it is tacitly unethical.
3. Sending a padded bill to a client is unethical and possible illegal.

Incidents like these are related to value systems, beliefs and culture. Laws are often created to combat unethical acts that threatens the society's survival. They are also used to reinforce existing systems to ethics. Codes of ethics also create a strong sense of professionalism. **A code of ethics** is a strong declaration of the principles and beliefs that govern how professionals or employees of a corporation or an association are expected to behave.

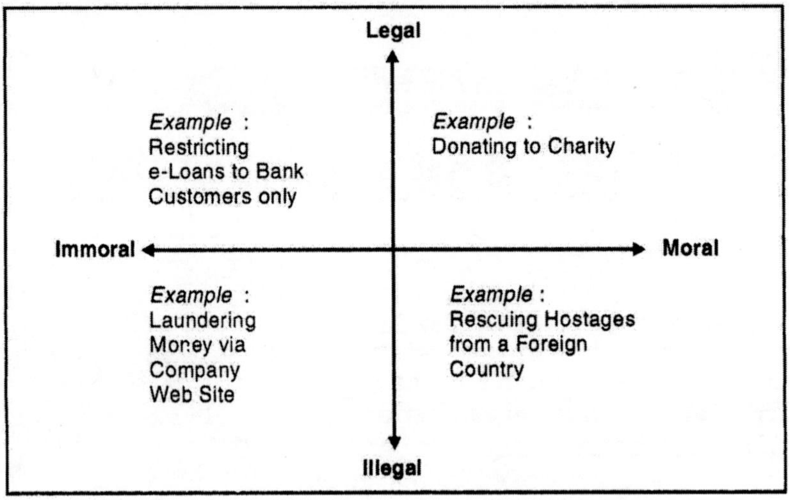

FIGURE 10.1. A model of acceptable behaviour.

10.1.2. Major Threats and Challenges to Social Ethics

Ethics in e-commerce is a serious concern today than ever in the past because the threats have steadily increased. Today's e-businesses face ethical dilemmas of dimensions not imagined 10 years ago. Technological advances have resulted in the need to revaluate ethical standard and their implications for privacy, confidentiality and integrity. Software copyright infringement, unauthorized e-mail access and the sale of competitive data are serious issues. High-speed, low-cost data transmission is raised new questions about property rights, piracy and plagiarism. All this is forcing a re-evaluating of the e-merchant's code of ethics. The main threats and challenges are :

1. Faster computers and more advanced networks
2. Sophisticated Global telecommunications
3. Massive distributed databases.
4. Ease of access to information and knowledge bases.
5. Transparency of software.
6. The idea that captured information can be used as a competitive weapon.

10.1.3. Improving Social Ethical Climate

E-business can take a number of steps to improve ethical behaviour in their IT departments, which is where websites are updated and programs are written :

1. To promote ethical behaviour throughout the organization, top manager should act as role models.

2. The company should establish a code of ethics that takes into consideration the state of technology (Intranets, Extranets, local area and wide area networks etc.). Goals should be realistic, achievable and agreed upon by all employees. Each organizational level should create its own customized ethics program, using the company's code of ethics as a framework.
3. Unethical behaviour should be dealt with promptly according to criteria and procedures set in advance.
4. The company should set up and support a string ethics training program for all new employees and reinforce the training on a regular basis.
5. The company should motivate employees to focus on honesty, integrity, fairness and justice as goals that are just as important as money or the bottom line.

10.1.4. Code of Social Ethics

The business ethics is a reflection of corporate culture and values, it means that the code of ethics should represent all that company stands for. The code should be all encompassing and stable over the time. To keep the ethical climate intact, an organization must stress regular self-assessment and encourage open dialogue within the organization.

10.2. PRIVACY ISSUE IN E-COMMERCE

Justice Brandeis vision of being "let alone" no longer suffices to define the concept of privacy in today's digital environment, where very bit of personal information can be transported and distributed around the world in seconds. Individuals should be able to interact in modern society without losing control over their personal information. The modern right to privacy also entails, therefore, the right to control our personal information even after we disclose it to others. Privacy means many things to many people and different things in different contexts. For the purpose of our discussion, we will examine several core "privacy expectations", that individuals have long held and which should carry over to their interactions on the Internet that are under siege.

It is one of the most pressing concerns of computer users and is inadequate addressed in e-commerce. E-companies are taking advantage of their ability to obtain information without the customer's knowledge or permission. Such private information is being documented, sold and used to promote e-business. Obviously e-business have an ethical responsibility to inform users of what information is being captured and how it is being used.

The thought of being watched is unsetting. Hidden video cameras, phone taps and surveillance bugs are all examples of technologies that are considered to be unethical (or sometimes illegal), since they allow data to be collected about individuals without their knowledge. There are websites whose only business is selling information about people who visit their sites.

10.3. E-COMMERCE LAWS

This section we will discuss range legal issues relevant to secure electronic commerce and, for the benefit of those readers without a legal background, explains the main legal principles and terminologies that need to be understood to fully appreciate the issues of EC. The issues covered include consideration of the state of evolving business practices, suggested solutions to some recognized problems and a status report on the evolution of relevant legal standards. It will be seen how the current state of the law presents some uncertainties in relation to the formation and enforcement of agreements. We have also discussed the recent efforts and measures to reduce these legal uncertainties, including domestic laws, international conventions, guidelines and model agreements and provisions between electronic commerce participants.

10.3.1. Legal Issues

It is true with any legitimate business, whether bricks or mortal or online, operates in a legal environment. Contracts, taxation and copyrights etc. are amongst legal issues of interest to all of us.

10.3.1.1. The Issue of Liability

The product liability makes a manufacturer liable if its product has a defective condition that makes it unreasonably dangerous to the user or consumer. Each entity involved in the process (e-merchant, vendor, shopper, manufacturer) is potentially vulnerable.

1. **Uniform Commercial Code** : A law drafted by the National Conference of Commissioners on Uniform State Laws, which governs commercial transactions.
2. **Warranty** : An assurance made by the seller about the goods sold.
3. **Express Warranty** : A warranty offered orally or in written by the maker of the product.
4. **Strict Liability** : A seller is liable for any defective or hazardous products that unduly threaten the user's safety.
5. **Implied Warranty** : A warranty that arises automatically from the fact that a sale has been made and the assumption that the product will do what it is supposed to do.
6. **Disclaimer** : Evidence of the seller's intention to protect the business from unwanted liability.

The Designer's Liability

The web designer is often responsible for system accuracy and reliability.

A variety of errors may become embedded in the system and these are :

(*i*) **Out-of-Bounds Error :** An error that occurs because either the software did not have the expertise to address the particular problem or the designer improperly condensed the technology.

(*ii*) **A Nontrivial Error :** A nontrivial error is one that triggers other areas in the software to malfunction and is difficult to correct. This type of error has a large financial impact on the e-business especially if the product is mass-marketed. The consequence of decommissioning the system or facing litigation.

As the designer relies on his or her experience to develop the product or software, when a malfunctioning occurs, the designer is vulnerable to charge of personal liability under the doctrine of *respondent superior* (an employer-employee relationship). If the designer is an employee of the organization that sells the software, the employing company is involved in the negligence action. In the end, the company is responsible for certifying the system before it is released for public use.

The User's Liability

Even end user of the product are not immune from lawsuits. They are directly responsible for proper use of the product. User's ability to refuse to comply with the product directions will increasingly come into question. By not properly using an available resource, user could be negligent by omission (what is called *passive negligence)*. For example, the use of an intelligent system in medical diagnosis could place the responsibility for using the system on the user as an *affirmative* duty.

10.3.1.2. Issues of Copyrights, Trademarks and Trade Names

The Internet copyright and trademark violations fall under intellectual property law. **Intellectual Property** includes software, music, videos, books, trademarks, copyrights and web pages. There is a growing controversy over, who, owns the intellectual property of domain names, programming and websites. Even the HTML coding of the website is in question.

Copyright is ownership of an original work created by the author. It is a form of intellectual property protection that covers the look, feel and content of printed media like articles and textbooks as well as software programs and software packages. **Copyright Law** gives the author or creator of a *tangible* product the right to prevent others from using the finished work. That is why authors and publishers place a copyright notice on the back of the title page. Copyright protection applies immediately upon creation of the manuscript.

A word, a picture or an image that identifies a product or service is intellectual property and is protected by a **trademark**. The term means registration of a company's trade name so that others cannot use it. It is also a word or a symbol that distinguish a good from other goods in the market. Who owns a trademark (or a copyright) is often a contractual matter.

10.3.1.3. Taxation Issues

One of the most controversial issues facing e-commerce in taxation, especially sales tax. In every state, a business enterprise is required to pay taxes and collect taxes. In businesses, computers and electronic machines compute and account for taxes to comply with local state and federal tax laws. On the Internet, tax collection is not so easy. What is collected depends on the location of the e-merchant's business, the location of the buyer, the types of good for sales, etc. Each state and country has different sales taxes and different jurisdictions. The rules for taxation also differ in different countries.

10.3.1.4. Consumer Issues

As electronic commerce continues to proliferate, consumers are becoming increasingly important participants. The law frequently provides added protections for consumers against fraud and unfair trade practices by unscrupulous merchants. The relative anonymity of parties in electronic commerce heightens the potential for such problems and the need for appropriate protection. Nonetheless, there has not been a major effort to reduce consumer risks in electronic commerce form the recognized consumer advocacy organizations as yet.

10.3.1.5. Intellectual Property Issues

Electronic commerce systems demand greater vigilance to ensure that parties do not violate or infringe upon existing copyrights, trademarks, patents or other intellectual property rights. In addition to copyright and trademark litigation regarding the content of certain communications, there has been significant legal activity in the trademark area with respect to the use of Internet domain names. Some companies have registered domain names using the trademarks or trade names of their competitors. Network Solutions Inc. (NSI), the entity that registers.com domain names, continues to struggle with increasing and novel disputes. In November 1996, an Internet International Ad hoc Committee (IAHC) was constituted to consider various reforms to the current Internet domain naming system.

10.3.1.6. Legal Disputes on the Internet

In B2C, several kinds of disputes have legal implications :

1. The customer pays for the merchandise, but the e-merchant fails to deliver.
2. The customer pays in full, but receive either the wrong merchandise or a partial order.
3. The customer does not like the product, but the e-merchant has no procedure for accepting returned merchandise.

4. The customer does not like the product, but the e-merchant refuses to accept returned merchandise or give credit to settle the dispute.
5. The e-merchant delivers but the costumer does not admit that he or she ever received the merchandise.
6. The e-merchant delivers, but the customer refuses to pay. His or her 14-years old ordered the product using a parent's VISA card without authorization.
7. The customer receive the merchandise, but it arrives damaged. The carrier denies responsibility, the e-merchant claims it is the carrier's responsibility and the vendor is located overseas with no customer service number.

The customer receives the merchandise but it does not operate properly. The e-merchant asks the customer to ship the product to the manufacturer at the customer's expense. The manufacturer has no in-house service center.

10.3.1.7. Issue of Cyber Jurisdiction

In terms of recourses, a lot depends on the laws protecting the buyer is the state in which he or she resides. It also depends on the place on operation of the e-merchant. Legal costs often exceed the value of the contested merchandise, which leaves the customer with no choice but to abandon the product. Where the web server or business is located determines the right of the customer to recover. Unfortunately, it is not always clear where the server is located, especially when the e-merchant has multiple web servers in different countries. What makes the situation more difficult is the fact that there is no legal restrictions on where the top-level domain (e.g., "UK") can be used and where it cannot be used. About the only thing that counts is the country where the e-merchant is located. That location determines the jurisdiction on the Internet.

One legal issue that has surfaced relates to products that are available in one country but are restricted in the country from which the customer is ordering. For example, in August 1999 Amazon.com was criticized by Germany's Simon Weisenthal Centre for selling books like *Mein Kampf* which are banned in Germany. While Amazon.com's German subsidiary does not offer these books on its web site, the U.S. website does. If the product is confiscated in Germany, what recourse does the customer have to recover the product ?

This matter boils down to the legal issue of **jurisdiction** or the legitimate scope of government power. A court must have jurisdiction over the litigants and the claims before it entertains a law suit. In the context of Internet commerce, this issue arises when there is a dispute between businesses from different states. For example, is a customer in Chicago required to travel to California to defend against a firm that is suing him for breaking a sales agreement ? Except in criminal cases, both state and federal law limit a court's jurisdiction over a defendant from another state. This means e-commerce and ensuring security and integrity in an e-business is still clouded by such legal issues.

10.3.1.8. Web Linking and Domain Name Disputes

The infrastructure of the Internet is designed around **hyperlinks**—text or image whose address can be linked to another web pages for reference. When you click on the link, it automatically goes to the attached location and the designated web page. This jumping from one site to another raises legal issues that include :

1. Referencing a linked site without permission from the site owner.
2. Retrieving or downloading information from a linked site without referencing or permission.
3. Unauthorized use of a company's registered trademarks.
4. Adding a web program to a company's web site without permission.

Here are some guidelines regarding domain names and trademarks :

(*a*) Find out whether the proposed domain name infringes on any trademarks. The fact that some other person or organization registers for a domain name does not give the owner the legal right to use it.

(*b*) Secure federal trademark registration for the proposed domain name. Once the name clears against possible claims of infringement, it should be registered as a trademark with the countries patent and trademark office.

(*c*) Registered the proposed domain name with interNIC (Internet Network Information Centre), the agency that represents the government in assigning domain names.

(*d*) In the event of a poached domain name, bring a lawsuit to force InterNIC to reassign the name to the original owner—the owner of the same name or trademark.

(*e*) Get permission before linking to other websites.

10.3.2. Forms of Agreements

Agreements play an important role in electronic commerce. In the absence of a rich complement of relevant law and trade practices including robust security standards, agreements increase the certainty of electronic commerce.

Agreements allow parties to benefit from freedom of contract. Without an agreement, the rights and obligations of parties engaged in electronic commerce will be determined (whether by design or default) by the prevailing law—which often differs from the parties' specific intentions and expectations. Thus, even when legal rules in the electronic commerce areas are established with greater certainty, parties can use agreements to deviate from these rules when such deviation is to their advantage. In other words, agreements provide an important opportunity for parties to structure electronic commerce relationships that are consistent with their precise business needs.

Agreements can be classified as either :

(1) Documenting a single transaction, such as the purchase and sale of goods, or

(2) Providing rules to facilitate electronic commerce, such as those diminishing electronic trade barriers caused by traditional paper and hand-written signature requirements. Such facilitating electronic commerce agreements include trading partner agreements, Value-Added Network (VAN) agreements, interconnection agreements and payment agreements. These types of agreements are discussed in this section.

One model of the relationships involved in some of these arrangements is presented in Figure 10.2.

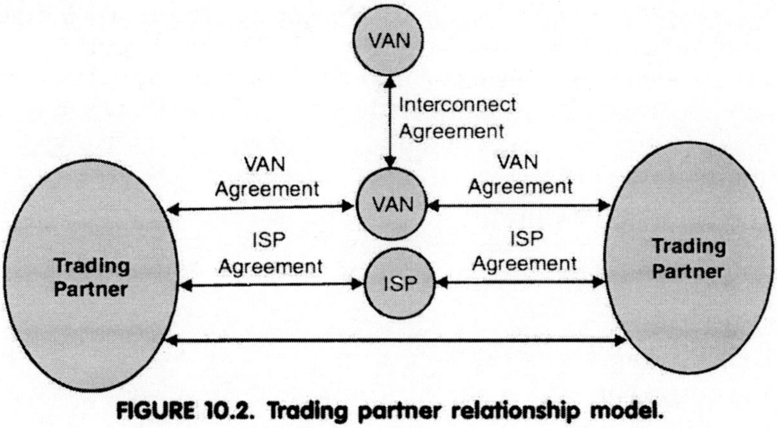

FIGURE 10.2. Trading partner relationship model.

10.3.2.1. Trading Partner Agreements

Trading Partner Agreements (TPAs) are agreements between two (bilateral) or more (multilateral) parties who wish to trade electronically. Such trading has traditionally been accomplished using EDI, but it now includes other electronic commerce methods for open systems such as the Internet.

10.3.2.2. Value-Added Network (VAN) Agreements

VAN agreements are executed between a value-added data communications service provider or VAN and a party desiring data communications services, typically for EDI. These bilateral agreements generally describe the data services to be provided, obligate the consumer to pay and include warranty disclaimers and liability limitations. VAN agreements may also include technical information regarding the use and processing of formatted message (such as EDI interchanges) and acknowledgments. These agreements also detail each party's responsibility for maintaining information security. The Internet Service Provider (ISP) agreements are a variant of the VAN agreements. VAN agreements typically address the following issues :

(*i*) Service description
(*ii*) Unauthorized access
(*iii*) Customer service
(*iv*) Ownership and use of customer data
(*v*) Data retention
(*vi*) Availability
(*vii*) Fees and surcharges
(*viii*) Termination of service
(*ix*) Third-party audits
(*x*) Processing schedules
(*xi*) Data recovery capabilities
(*xii*) Dispute resolution
(*xiii*) Apportionment of liability

10.3.2.3. Interconnection Agreements

To facilitate reliable communications between trading partners connected to different VANs, many VANs interconnect their systems or networks. VANs execute interconnection agreements with other VANs to state how their interconnections will operate and to protect themselves from liability. The precise terms of an interconnection agreement are a function of the parties' respective bargaining power, business needs and technical protocols. These protocols vary in their capacity to communicate acknowledgments, transmit billing information and provide security. Interconnect agreements, typically apportion liability between the VANs, indicate how and when data are to be exchanged and facilitate inter-VAN communications generally. Interconnection agreements typically include the following provisions.

(*i*) Message transfer responsibilities
(*ii*) Acknowledgments
(*iii*) Error resolution
(*iv*) Level of service
(*v*) Billing/reimbursement practices
(*vi*) Charges and settlement
(*vii*) Issue of liability

10.3.2.4. Payments Agreements

Payment agreements enable or enhance the certainty of success in making payments electronically. Payment agreements take various forms, including

(*i*) Financial EDI (or electronic commerce payments) agreements

(*ii*) Credit card and Internet payment service provider agreements

(*iii*) Electronic Funds Transfer (EFT) agreements

10.3.2.5. Security Provision in Model Agreements

Since treatment of security in both domestic and international electronic commerce agreements has traditionally been quite modest, this chapter provides special focus on agreement provisions and related infrastructure supporting secure electronic commerce. Examples of the information security provisions included within widely-recognized model agreements are:

Model EDI Trading Partner Agreement : "Each party shall use those security procedures properly, which are reasonably sufficient to ensure that all transmissions of Documents are authorized and to protect its business records and data from improper access".

Model Electronic Payments Agreement : "Each party shall employ reasonable security procedures to ensure that Transaction Sets, notices and other information specified in this Agreements that are electronically created, communicated, processed, stored, retained or retrieved are authentic, accurate, reliable, complete [and confidential]".

European Model EDI Agreement : "The parties undertake to implement and maintain control and security procedures and measures necessary to ensure the protection of messages against the risk of unauthorized access, alteration, loss or destruction".

10.3.3. Encryption and Decryption Laws

Encryption is not a pleasant word to use in certain countries. Some Middle East Countries, for example, prohibit any form of Encryption for business or personal use within the country or across the border. Obviously, encryption poses a threat to the powers of many government, but because of the impressive surge of traffic on the Internet, awareness of security has increased significantly within governments worldwide.

Cryptography has had its share of attention over the past decade. Among the issue are these: what can be exported and what cannot be exported ? How safe is the computer from Internet crime ? In the United States, there tends to be a difference of opinion on encryption between federal agencies like the FBI and big business. Secure electronic payments require secure lines. There is a belief that encryption makes line more secure.

Electronic commerce is a new technology, which is growing rapidly. It has the ability to create a truly global digital economy, but at present current legislation does not encourage the uptake of this technology. The nature of the Internet and the globalization of the world economy mean that developments in electronic commerce create legal problems concerning :

1. Legal Recognition of Electronic Messages.
2. Contracting by Electronic Means.
3. Jurisdiction.
4. Evidence.
5. Consumer protection and law enforcement.
6. Intellectual property.
7. Payment systems, electronic money and electronic banking.

8. IT Security (computer and internet crimes)
9. Liability of online intermediaries.
10. Infrastructure legal regulation.
11. Taxation.
12. Customs.

10.3.4. United Nations on E-Commerce

The United Nations Commission on International Trade Law (UNCITRAL) is attempting to develop uniform international rules that would validate and encourage the use of EDI. In 1995, UNCITRAL adopted the draft Model Law on Legal Aspects of Electronic Data Interchange (EDI) and Related Means of Communication. The Model Law is intended to serve as a model to countries in order to create uniform law and practice involving the use of computerized systems in international trade. The objectives of the Model Law are essential to improve the efficiency in international trade since it will, if adopted by a large number of countries, enable and facilitate the use of EDI and the related means of communication and providing equal treatment to users of paper-based documentation and to users of computer-based information. The Model Law was finalized and approved in the UNCITRAL twenty-ninth session, 28 May to 14 June 1996. The Model Law is before the United Nations General Assembly and is currently awaiting approval.

The Model Law applies to any kind of information that is transferred in the form of a data message used in commercial activities. Ocean bills of lading are one kind of document within the scope of the Model Law. However, the legal aspects of bills of lading are key issues of the Model Law developments, as noted in the UNCITRAL Working Group on EDI of its thirtieth session, 26 February - 8 March 1996. This reported that future work could focus on EDI transport documents, with particular emphasis on maritime electronic bills of lading and the possibility of their use in the context of the existing national and international legislation dealing with maritime transport.

It should be noted, however, that the Model Law has no legal force of its own. Its provisions will have legal enforcement only if they are enacted in national law. For this reason UNCITRAL also adopted a Guide to enactment of the Model Law.

10.3.5. International Chamber of Commerce (ICC)

The ICC has published the General Usage in International Digitally Ensured Commerce (Guide), formerly known as Uniform International Authentication and Certification Practices (UIACP). Guide "aims to draw together the key elements involved in electronic commerce, to serve as an indicator of terms and an exposition of the general background to the issue. It also addresses one of the key problems in talking about electronically signed messages, in that they are not signed physically, but require the intervention of an electronic medium. This in turn alters the function of the signer, and introduces problems that a physical signature does not encounter, most especially the possibility of use of the medium by a third party. The GUIDEC therefore adopts a specific term, "ensure", to describe what elsewhere is called a "digital signature" or "authentication", in an attempt to remove the element of ambiguity inherent to other terms employed."

10.3.6. E-Commerce Laws in India

The Parliament of India passed its cyberlaw in the form of the Information Technology Act, 2000, which provides the legal infrastructure for e-commerce. The Act received the assent of the President of India and became the law of the land on 17 October 2000.

The objective of the Information Technology Act, 2000 would be to provide legal recognition for transactions carried out by means of electronic data interchange and other means of electronic communication, commonly referred to as electronic methods of communication and storage of information. The act would also facilitate electronic filing of documents with various government agencies and further to amend the Indian Penal Code, the Indian Evidence Act, 1872, the Banker's Book Evidence Act, 1891 and the Reserve Bank of India Act, 1934 for related matters.

The Act thereafter stipulates numerous provisions in order to provide for the legal framework so that legal sanctity is accorded to all electronic records and other activities carried out by electronic means. The Act further states that unless otherwise agreed to the acceptance of a contract expressed by electronic means of communication shall have legal validity and enforceability. The Act would facilitate electronic intercourse in trade and commerce, eliminate barriers and obstacles to electronic commerce that result from the celebrated uncertainties relating to writing and signature requirements over the Internet. The objectives of the Act also aim to promote and develop the legal and business infrastructure necessary for implementing electronic commerce.

Chapter II of the Act stipulates that any subscriber may authenticate an electronic record by affixing his digital signature. It further states that any person can verify the electronic record by the use of a public key of the subscriber.

Chapter III contains details about e-goverence and provides, among other things, that where any law provides that information or other matters shall be in writing, typewritten or printed form, then, notwithstanding anything contained in such a law, that requirement should be satisfied if the information or matter is :

(a) Rendered or made available

(b) Accessible to make it usable for subsequent reference.

That chapter also provides details about the legal recognition of digital signatures. The various provisions give further elaboration about the use of electronic records and digital signatures in government agencies. The Act also refers to publication of rules and regulations in an Electronic Gazette.

Chapter IV gives a scheme for the regulation of certifying authorities. The Act provides for a controller of certifying authorities. Who shall perform the function of supervising the activities of certifying authorities as well as setting standards and conditions governing the certifying authorities. The controller also specifies the various forms and the content of digital signature certificates. The Act acknowledges the need to recognize foreign certifying authorities and it further details the various provisions for granting the license to issue digital signature certificates. The duties of subscribers are also covered. The Act also covers penalties and adjudication for various types of offences and mentions the power and qualifications for the adjudicating officer.

A provision in Chapter X foresees a Cyber-Regulations Appellate Tribunal where appeals against the orders passed by Adjudicating Officers could be referred. The tribunal would not be bound by the principles of the Code of Civil Procidure, but would follow the principles of natural justice and have the same powers as a civil court. Any appeal against an order or decision of the Cyber Regulations Appellate Tribunal would be made to the High Court.

Chapter XI covers various offences and stipulates that the investigation must be by a police officer only, and that officer should have the rank of deputy superintendent of police or higher. These offences include tampering with computer source documents, publishing obscene information in electronic form, breach of confidentiality and privacy, misrepresentation, publishing a digital signature certificate that is false in certain particulars and publication for fraudulent purposes.

Hacking and penalties if found guilty have been defined in Section 66. For the first time, punishment for hacking has been designated as a cyber crime.

The Act also provides for constituting the Cyber-Regulations Advisory Committee, which would advise the government about any rules or other matter connected with the Act. The Act also has four schedules which amend the Indian Penal Code, 1860, the Indian Evidence Act, 1872, The Bankers' Books Evidence Act, 1891. The Reserve Bank of India Act, 1934 to make them conform with provisions of the IT Act.

Overall, the Information Technology Act, 2000 is considered to be a commendable effort by the government to create the necessary legal infrastructure to promote and encourage the growth of electronic commerce.

10.4. E-COMMERCE: GOVERNMENT'S POLICIES AND AGENDA

There is always a close relationship between the Government and the Industry. The Government and the Industry should work together in order to have a stable economy. The Government can play a lead role in providing support for the implementation of any new technology in terms of infrastructure and other related issues. Here some such factors have been highlighted which show how the government can play an important role in implementing Electronic Commerce.

Governments have a direct stake in Electronic Commerce by virtue of the fact that in virtually every country, the public sector is the largest single procurer of goods and services. Increasingly, moreover, the general trend in most countries is for government to seek private sector sources for as many products and services as possible.

As governments seek to increase their own implementations of Electronic Commerce solutions, they become subject to the same kinds of pressures for organizational and managerial change as do private firms. In government, the concept of Electronic administration is evolving in a somewhat symbiotic relationship with the development of Electronic Commerce. Together, these forces are bringing many of the interests of government and industry closer together, both with respect to the benefits and the problems of Electronic Commerce.

10.4.1. Role of Governments to Promote Electronic Commerce

Governments can play two positive roles in the development of Electronic Commerce :

(1) Direct promotion through the application of Electronic Commerce principles in government administration and procurement and in the provision of public services.

(2) Facilitation of the development Electronic Commerce through the provision of a legal, regulatory and infrastructure environment that encourages the development of Electronic Commerce.

In countries where the private sector has already taken lead in Electronic Commerce, government can support these initiatives by providing a supportive legal and regulatory environment and by ensuring that government administrative practices are sensitive to the characteristics of the electronic marketplace. In the case of newly industrializing and developing countries, governments should be encouraged to take a creative role in promoting Electronic Commerce. In this context, public interventions of various kinds can act as important catalysts for the diffusion of Electronic Commerce applications in the private sector.

10.4.2. Defining the Policy and Regulation for Electronic Commerce

In the modern era, governments have exercised controls over the commercial environment for two principal reasons.

Firstly, governments have sought to prevent abuses of the market system through such measures as competition policy and the regulation of financial institutions. Secondly, governments have sought to promote commercial environments aimed at yielding particular social and economic outcomes, using such measures as foreign investments and exchange policies, consumer protection legislations or skills and training programmes. In many cases, initiatives to impose controls have come from business and industry as well as from government itself and the original objectives of most of these controls was to institutionalize conditions of trust between buyers and sellers.

However, most of the legal and regulatory mechanisms being currently applied by governments to commercial activity were communication systems. The majority of these mechanisms are national in orientation.

Electronic Commerce has to ignore sector and national boundaries and it tends to accentuate the intangible aspects of commerce over the tangible aspects. This presents a policy challenge for governors and governed, alike.

The entire basis of commercial policy making is now coming under close scrutiny. Electronic Commerce presents many fundamental challenges to the policy framework, especially where the enforcement of legal and regulatory provisions is concerned. Governance of the electronic marketplace is certainly beyond the scope of action for individual governments, requiring instead an increased emphasis on intergovernmental cooperation.

Electronic Commerce provides governments with the opportunity to re-examine the established of policies for trade and commerce. Many major kinds of strategic decisions about the national and international provisions of network services now reside almost exclusively with the private sector. At the present time, many Electronic Commerce developments are occurring virtually outside the realm of government regulation.

10.4.3. Policy Recommendations for E-Commerce Growth

In order to grasp the developing opportunities in Electronic Commerce creatively, efficiently and productively, we consider the priority areas for government action.

10.4.3.1. Supporting Opportunities for the Growth of E-Commerce

Government can support the e-commerce by providing following:

(a) Regulation of the Information Infrastructure

Electronic services infrastructures like telecommunications, television and data networking, as well as the regulatory regimes that effect them, must be permitted and encouraged to converge in order to reflect that rapid convergence of networking technologies employed in E-Commerce applications.

(b) Standardization

Although in many cases non-proprietary standards are the preferred solution for E-Commerce applications, the rapid pace of technological and business development will certainly outstrip the ability of standards organizations to respond to market needs in a timely manner. We urge governments to adopt a pragmatic approach that does not discourage the development of widely accepted proprietary solutions becoming adopted as if they were standards for Electronic Commerce.

(c) Secure Information Exchange

While recognizing the legitimate security and law enforcement responsibilities of government, it is essential that they be prepared to accept multiple solutions to encryption problems.

National data protection laws should be harmonized early in order to prevent situations that could destroy public confidence in Electronic Commerce where privacy and confidentiality implications are concerned. Governments should not act to restrict transborder data flows unduly, but should act multilaterally to minimize the possibility of abuse in the utilization of customer information when databases are transported form one jurisdiction to another.

(d) The 'Physical' Infrastructure

Where Electronic Commerce is applied to the production and distribution of tangible goods, a close relationship exists between the electronic and the non-electronic components of the commercial infrastructure. Electronic Commerce cannot function on its own. Costly and inefficient systems for the physical transportation of goods could slow down the development of the electronic marketplace. Governments have a major role to play in ensuring that physical transportation infrastructures are developed in coherent, coordinated ways, such that they complement the transaction efficiencies generated by Electronic Commerce and do not interfere with speedy process.

(e) Proper Action Plans to Reflect the Pace of Technical Change

In order to ensure close communication with industry, governments should give early consideration to the development of high-level statements outlining proposed actions of relevance to Electronic Commerce, along with time scales.

10.4.3.2. Promoting Productive Industry/Government Relations

Government can play a constructive role to stream line the relations with industry as follows :

(a) Coordinating Public and Private Sector Electronic Commerce Activities

To ensure the productive coordination of public and private sector agendas for the application of Electronic Commerce, every government should have a Chief Information Officer for national information and communication activities. In general, all government actions in areas of importance to Electronic Commerce should be given high visibility in order to generate and sustain public confidence and trust in the emerging Electronic Commerce environment.

(b) Flexibility in Public/Private Sector Relationships

Industry and government should maintain a dynamic international dialogue on the harmonization of Electronic Commerce and Electronic Administration principles. Public sector practices must be made responsive to the substantial changes in the business must be made responsive to the substantial changes in the business environment that are excepted to occur with the diffusion of Electronic Commerce. This applies particularly to procurement practices and to all administrative reporting procedures that are required of industry by governments.

(c) Awareness, Education and Skills

Governments should promote awareness of the potential of Electronic Commerce in all sectors of the economy and, in consultation with industry, work to reflect the rapidly changing professional and skills requirements of the electronic marketplace in education and employment programmers.

10.4.3.3. Defining New Rules and Regulations for an Electronic Environment

(a) Reforming Regulatory Practices

Government support for the concept of liberalized electronic services markets must be backed up with real changes in regulatory practices. Badly conceived deregulation and reregulation can provide new forms of protection to existing monopolies and can be worse than preserving the status quo.

New kinds of regulation may be required to prevent new infrastructure providers from acquiring market power on a scale equivalent to that exercised by incumbent infrastructure monopolies, but new regulation must not prevent better infrastructure management.

(b) Clarification and Revision of Laws and Regulations

As a matter of urgency, governments need to clarify the legal definitions, practices and structures that pertain to commercial activities in an electronic environment and to seek multilateral agreement on critical legal matters, especially the laws regarding residency, agency, liability, auditability, control of databases, unauthorized use of databases and data protection.

Where appropriate, governments should adjust existing laws and regulations so that they apply to 'intangible' as well as 'material' product environments. They should ensure that all future actions regarding consumer protection laws and regulations are closely coordinated with developments in Electronic Commerce.

(c) Enforcement of Laws

An internationally clarified legal environment must be supported by international agreement on policing and enforcement.

(d) Protection of Intellectual Property

The scope, nature and international coherence of IPR systems must be reviewed in light of emerging Electronic Commerce requirements.

SUMMARY

E-commerce has its own issues like social issue and privacy issue. Certain rules and regulations are required to perform the e-commerce operations. The government of any country plays major role for the success of e-commerce since it is the government, which decides rules, laws and policies for the e-commerce implementation and operations.

REVIEW QUESTIONS

1. How E-commerce can impact on society ?
2. What are the various social issues in E-commerce ?
3. Discuss the various issue to implement the E-commerce ?
4. What is the rate of forms of agreement in E-commerce ?
5. What is the importance of security provisions in E-commerce ?
6. What is role of United Nations in E-commerce ?
7. How a national government can play a role to decide the policies and agenda for E-commerce ?
8. What is the rules and regulations of E-commerce in India ?

❑❑❑

11

CHAPTER

IMPLEMENTATION OF E-COMMERCE

INSIDE THIS CHAPTER

11.1. Designing of a Well-Functioning Commercial Website
11.2. Developing E-Commerce Enabled Application

11.1. DESIGNING OF A WELL-FUNCTIONING COMMERCIAL WEBSITE

A well-organized website should be easy for a customer to use and easy for the owner to maintain. It takes a lot to build a well-organized website. Below is a list of the most important elements that a good website should have:

1. **A Well-Organized Collection of Products and/or Services :** To maintain a website, the owner should be able to change product information easily. A website where a price change of a single product requires changing 3 web pages is just bound to have inconsistencies. The owner should be able to add or remove products based on every day's availability, change prices, add product cross-references, etc. without making the website inconsistent even for a minute.
2. **A Convenient Way for a Customer to Select Products :** Usually implemented as a shopping cart or a shopping basket. The customer should be able to select and delete products while browsing the website.
3. **Convenient Order Forms :** The form should be flexible enough to allow to specify a different address for the product delivery, a gift message, etc. It should have as few required fields as possible. A returning customer should be provided default information so that not to type it every time. For both the customer and the owner, it is essential that the form catches simple types.
4. **Convenient Ways of Payment :** There should be options of paying by a credit card, by a check (not everyone has credit cards!), and by a credit card over the phone if the customer is not comfortable sending his/her credit card number. The options may include some electronic payment systems. For the owner, there should be a quick way to verify the credit card information or in some other way to check that the payment is valid.

5. **Secure Communication System** : Not only to protect transmission of a credit card number, but also to guarantee privacy of the customer (including details of the purchase). A website might have a user registration system with a password, in which case all transactions by the user should be private. It is also important to prevent unauthorized access to the website (by a hacker or accidental).
6. **Some Way of Storing Information about Customers :** This is convenient for customers so that they don't need to reenter their information every time they access the site. It also allows to "customize" the website for someone's interests. This can be done via customer registration or by means of "cookies". The owner can benefit a lot from storing information about a customer: he/she can customize ads based on the customer's profile or send an e-mail advertising a new product (but keep in mind that many people don't like this!). However, the greatest benefit is the owner's ability to monitor customer's behaviour: which pages have the customer visited and which purchases (if any) he/she have made afterwards.
7. **A Way of Keeping Information about Orders :** This allows customers to track their orders, and for owners to get all kinds of financial and statistical information. It is also important to keep order information in case of later disputes.
8. The last, but not the least, **customer support and feedback**. There should be online documentation for all products ever sold on the website, various FAQs, and, ideally, a way of customers to post their opinion about the product. Easy access to this information may make a difference between a frequently visited website and a lonely online looser. This aspect of a website cannot be completely mechanized: a human being has to answer e-mail, judge the relevance of customer's comments (otherwise the website may be flooded with nonsense), organize comments by topics, and so on. However, there is a lot that can be computer-aided in this process, for instance sorting incoming messages by their title and/or return address to forward them to an appropriate customer support person.

Note that we are only considering business-to-customer interactions here (so-called **B2C**), leaving out all the features related to business-to-business (**B2B**) interactions. While such interactions are basically similar to B2C, they can be optimized if a company has stable business partners which have more powerful computers than an average customer does. Even a customer oriented website has a B2B part when dealing with vendors of products. We are not considering B2B connections in designing a website.

11.2. DEVELOPING E-COMMERCE ENABLED APPLICATION

Here we will study the steps involved in setting up e-commerce and requirements for our online web store.The working of ecommerce has been shown in Figure 11.1.

The consumer first moves through the internet to the merchant's website. At the website, the consumer is briefly given an introduction to the product or services the merchant offers. It is at this point that the consumer makes the decision to visit the web store by clicking on a link or button located on the web page (e.g., Buy Now, Shop Online, or an image of a shopping cart button are common entry points into a web store). After choosing to visit the web store, the consumer is typically connected to an online transaction server located somewhere else on the internet which runs software commonly referred to as a shopping cart application. The shopping cart application has been setup by the merchant to display all products and services offered, as well as calculate pricing, taxes, shipping charges, etc.

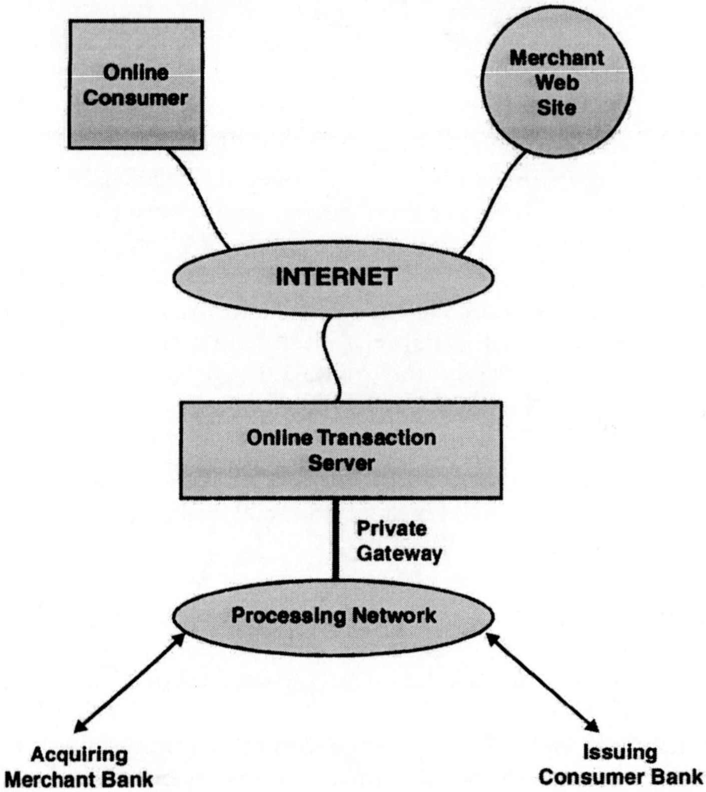

FIGURE 11.1. Working of e-eommerce.

From there, the consumer decides that he wants to purchase something, so he enters all pertinent information, provides a credit card number, and is immediately notified if the order was successful.

Now, there are several basic steps we will need to accomplish before becoming Commerce Enabled.

1. Getting an Internet
2. Getting a Merchant Bank Account
3. Obtaining a Web Hosting Account
4. Web Design Considerations
5. Registering a Domain Name
6. Obtaining a Digital Certificate
7. Finding a provider of online transactions
8. Finding shopping cart software

11.2.1. Getting an Internet Connection

Getting an Internet connection is not a big deal nowadays. We can get Internet connection from local ISP (internet service provider company).

We can get Internet connection by various ways like :

(*i*) Cable line
(*ii*) Lease line
(*iii*) Telephone line
(*iv*) Satellite link

In india many ISP like Satyam, BSNL, MTNL, Bharti Telecom, Reliance Infocom, Mantra-Online and Tata etc. provides intarnet connection.

11.2.2. Getting a Merchant Bank Account

In order to be able to accept credit cards, we must apply for an account with a credit card merchant account provider. This can be relatively easy or somewhat difficult, depending on which country we live in, and the type of business we are running. In the past, many businesses would sign up for credit card processing through their own commercial banking institutions. However, the card processor you get signed up with must be able to support internet-based transactions in order for e-commerce to work.

Many card processors treat e-commerce as separate types of transactions and may require us to open a separate account to segregate card present and non-card present transactions.

If we plan on using ManageMore's eConnect Cart, we may want to contact Intellisoft to ensure that the Merchant Account we plan on using will work seamlessly with the shopping cart software as well as within ManageMore.

11.2.2.1. Merchant Account versus Bank Accounts

In our day to day activities of depositing money into our bank account we physically go to the local branch or visit a remote ATM (automatic teller machine) and perform the transaction.

A merchant account is different, because we can't be physically present during the 'virtual' transaction. The merchant account is merely a specialized bank account where money from credit card sales is first routed to and held before transfer to your own checking/savings account. When a payment has been accepted from a client online you will need a 'third party' to approve the transaction, hold the money for us, and finally deposit it into our account. That is the purpose of a merchant account.

This process is usually fully automated in e-commerce transactions through a payment gateway. A payment gateway is the software programming term used by merchant account providers that represents the front door to their credit card approval programs. Since our shopping cart is a software program it needs a place to 'talk to' when requesting the approval of a credit card transaction. The shopping cart uses the payment gateway for such purposes. It contacts the payment gateway of the merchant account over the web, sends the payment amount and credit card details and then waits for an approval or denial message. Depending upon what message is received will dictate if the transaction can be completed and what the shopping cart should do next. If it was successful (the credit card was valid, the expiration date valid, and there was room on the card for this amount) then the money will transferred into our business account in real-time, or during various points in a 24 hour period.

Merchant accounts have different policies, different pricing's and different types of payment gateways that are written in many programming languages. When selecting a merchant account it is very important to understand the following :

1. How many transactions a month do you expect to process.
2. What is the average size of a typical transaction.
3. What programming language is the shopping cart written in (*i.e.* Java, ASP, .NET, PHP, PERL etc.).
4. What is the operating system of the web server (*i.e.* Windows, Linux, etc.) .

This portion of the e-commerce setup is the most complicated because it typically requires knowledge and expertise in web technologies beyond the scope of most new merchants. It is suggested that we consult a qualified e-commerce web development company or consultant who can further assist us.

11.2.3. Web Hosting

Web hosting is a very important step in this process, as this is how we gain a presence on the Internet in the first place. There are actually two scenarios that can be used for web hosting. Scenario 1 involves setting up and maintaining our own web server, while Scenario 2 involves outsourcing all web hosting administration to an ISP.

An Internet Service Provider (ISP) is a company that provides us with internet access and limited hard drive space on their web servers for hosting our website. We will need to setup an account for internet access with the ISP of our choice.

The following should be noted when searching for an Internet Service Provider :

1. Choose an ISP that is known for good technical support and has knowledgeable people familiar with e-commerce sites.
2. Choose an ISP that consistently has fast connection speeds and few interruptions.
3. As with any company we do business with, make sure the ISP is reputable.

11.2.4. Registering a Domain Name

If our business already has a corporate website implemented, then we probably already have a domain name and don't need to go through this process.

Domain names are the names for computers on the Internet that correspond to IP (Internet protocol) numbers to route information to addresses on the Internet network. Domain names serve as a convenient way of locating information and people on the Internet. In layman terms, will it be important to you, for customers to find your website by typing "123.123.456.456" or by typing something simple to remember like "www.kiet.edu?"

Registering a domain name is one of the most important decisions we can make for our online identity. Our domain name says who we are to our clients, our peers—the whole world.

The basics for registering a domain name are :

1. Contact a domain name registrar on the internet to register for a domain name. There are many to choose from, just do a web search on "domain name registrar" to get us started.
2. Select a unique domain name we would like others to use for finding our website. One place to go for checking availability of a domain name is **www.whois.net**
3. Expect a registration fee of \$5-\$25 annually for the central registrar to keep your domain name active on the internet.

There are many other questions that arise when considering a domain name for our business that go outside the scope of this book. For more FAQ's on domain names, check out **www.domainbank.net** or **www.netsol.com.**

11.2.5. Obtaining a Digital Certificate

A digital certificate, also known as a SSL Server Certificate, enables SSL (Secure Socket Layer encryption) on the web server. SSL protects communications so we can take credit card orders securely and ensure that hackers cannot hack our system. Any e-commerce company that provides us with an online web store will require us to have SSL before we can use their services. Thankfully, for most people obtaining a digital certificate is not a problem. For a minimal fee, one can usually use the certificate owned by the web hosting company where your page resides. If we are a larger company, however, we may want to get our own digital certificate. A certificate can cost anywhere from \$30 to \$125.00 annually.

11.2.5.1. Signing Digital Certificates

After we created a digital certificate, we want to have it signed by its owner. After the digital certificate has been cryptographically signed by its owner, it is difficult for anyone else to forge. For sites involved in e-commerce or any other business transaction in which authentication of identity is important, a certificate can be purchased from a well-known certificate authority such as VeriSign or Thawte.

11.2.6. Finding a Provider of Online Transactions

To process online orders, we must offer online payment options. The most widely used form of payment currently is the credit card. Marketing studies show that we'll lose 60 percent to 80 percent of our potential orders if our Website is not set up to accept credit cards; they also show that if we offer credit card payment, not only will you receive more orders, but those orders will be substantially larger.

Credit cards enable impulse buying, reassure customers of our legitimacy, and simplify our billing. Other methods of collecting payment are becoming available and include charging purchases to a phone bill, using electronic funds transfers (EFT), paying by electronic check and various forms of prepayment. Each of these methods requires payment processing either in the form of software added to our Website or by linking to a payment processing service.

Step 1 : Evaluate Alternative Online Payment Methods : Credit cards still reign as the leading method of payment for online purchases, but other payment options are available. Our product and our customers' buying preferences will influence which payment methods you accept. In other countries, credit cards are not as important, so you may want to consider offering alternatives for our international customers.

Offering multiple payment options on our Website, if we can afford it and maintain our profit margin, is a means to increase sales by increasing customer convenience and confidence. Many alternative methods are better suited to micro-payments, charges under $1, because the processing costs are often lower and credit card merchant-account fees don't apply.

Step 2 : Determine the Fee Structure that Maximizes Your Profit Margin : Not every product sells the same way, and not every merchant account provider charges us the same way. Choose a provider that suits our business. Begin by considering the nature of the products we sell are they large and expensive? Perhaps then we ought to seek a MAP that offers a higher flat-rate transaction fee and minimizes the discount rate, since even a hefty Rs. 100 transaction fee will be far lower than a 2.5 percent deduction from the charge. On the other hand, if you rely on small, high-volume sales, even a Rs. 25 transaction fee can erase our profits.

Step 3 : Specify Your Technical Requirements : Different MAPs require different "gateways" on our site. These gateways are the pieces of code that transmit our customers' orders to and from our bank's transaction authorizing agent. If we plan to manually process our orders, a secure Web form might be good enough to capture credit card information that we can process offline.

Step 4 : Evaluate Your Business's Credit-Worthiness : MAPs, like most banks, pay close attention to the companies with which they do business. Such factors as our company's length of time in business, outstanding debt, debt payment history, goods and services offered and even our personal history (for new businesses) will affect the fees our company pays to process credit card transactions on the Net.

Step 5 : Find MAP (Merchant Account Providers) We can Work with : Many merchant account providers refuse accounts to start-up firms or firms and individuals with bad credit histories. Some MAPs will not accept "high-risk" accounts, a term that usually encompasses adult sites, online casinos, and sites operated by firms outside the MAP's own nation. Others refuse to process any transactions that originate on the Internet—even from their own existing brick-and-mortar clients—or may require that we create a separate merchant account to process orders that are not taken face-to-face but are received by mail, phone, or via the Internet.

Step 6 : Compare Fees and Technical Capabilities : Once we've developed a list of merchant account providers who might offer us an account, we need to compare the different MAP offerings. Be certain to ask detailed questions about each MAP's technical requirements, and make sure our system can work with our MAP's gateways—the software that actually submits our customers' credit card information for payment authorization.

Step 7 : Minimize Credit Card Chargebacks : Discussion about consumer credit card protection for Internet purchases has become intense. But the fact is that U.S. federal law limits a consumer's liability for unauthorized charges to $50, whether the purchase was made face-to-face or on the Internet. No such protective legislation exists for merchants, however, and they bear the full cost of fraudulent charges as charge backs from their banks.

When a fraudulent credit card transaction takes place without the physical card being presented to the merchant, or funds are uncollectible for some other reason, merchants are charged the sale amount by the cardholder's bank. This is known in the industry as a charge back. Merchants may also be asked to pay penalty fees in addition to the cost of the original charge. Though it has not been as hot a topic for e-tailers, credit card charge backs pose a serious threat to profits. To reduce revenue losses due to credit card fraud, online businesses need to take steps to reduce the risk they take with every order received through their Websites. Find out how credit card charge backs occur and what we can do to protect ourself.

Step 8 : Secure all Our Transaction Data and Prevent Fraud : Credit card information is extremely sensitive, and plenty of villains are waiting to exploit any breach in our security. Additionally, online merchants are as susceptible to credit-card fraud as face-to-face retailers. Make sure our merchant account provider has addressed these issues.

Step 9 : Prepare for International Payment Processing : Many payment processing and merchant account providers do not accommodate international commerce. If we plan to market our product globally, we may need to search specifically for an international provider.

11.2.7. Finding Shopping Cart Software

The next step up is to use a service like Yahoo Stores. Here we can purchase a fully functioning shopping cart for a monthly fee. A merchant account is not required if using paypal or the service providers own payment processing service. If we do have a merchant account it most likely can integrate into this shopping cart system. The shopping cart provided typically is template based which will give us some flexibility in the look and feel of our website. They offer a lot of features both to assist in selling our products and also to manage the sale. These come standard with most packages.

The biggest advantage of this service is it is practically hands free. The web hosting, the shopping cart, the merchant account and the tools are all included in the monthly fee. The biggest disadvantage is that there is no ability to customize the software, add or modify the features, or create a unique and custom website that will differentiate you from our competitor using the same or like service.

This service is a great solution for anyone that has had success with paypal or wants to try out a shopping cart without making a big up front investment.

11.2.7.1. Shopping Cart Software

The last option is the most complex. This option offers the most flexibility in creating a unique shopping experience for our buyer. In this case we will need to acquire a web host our website. Establish credit, open a merchant account, and purchase a digital certificate to ensure encrypted transactions for our buyers. Lastly we will need to contract out an experienced ecommerce development firm to create a custom shopping cart for our exact needs. This shopping cart will have custom graphics, and text that will create a polished look to our product line.

The biggest disadvantage in this case is the up front cost. These costs can range from several thousands for small ecommerce sites all the way up to hundreds of thousands for professional retail stores with thousands of products. The biggest advantage however, which often outweighs the other choices is the clean, polished, unique look and feel. The creation of a unique shopping experience that rivals the large retail stores and that of our competitors.

This option is best for the serious retailer or the small shop looking to create a strong name for themselves in the market place.

SUMMARY

After studying e-commerce concepts on previous chapters, we have seen implementation of e-commerce business in this chapter. The basic steps are getting internet, a merchant bank account, obtaining a web hosting account, consideration about web design, registering a domain name, obtaining a digital certificate, finding a provider of online transactions and finding a shopping cart software.

REVIEW QUESTIONS

1. Discuss the most important elements that a good website should have.
2. What are the various steps, we will need to accomplish before becoming e-commerce enabled ?
3. What is merchant account and how can we get it ?
4. Why should be go for domain name registration ?
5. How should we select a provider of online transactions.

❑❑❑

APPENDICES

APPENDICES : PRACTICE TUTORIALS

Q. 1. Fill In The Blanks.

(*a*) The Information Superhighway is also called as __________ ?

(*b*) The main service which is used in E- commerce for data transmission is _________ ?

(*c*) E-commerce uses _________ as a network through which a number of organizations are interact to each other.

(*d*) Three basic requirements of a market 1. ______, 2.______ and 3.______ to setup a business.

(*e*) The different online methodology (ways) which used in E-commerce are _________.

Ans. (*a*) I-Way

(*b*) Electronic data interchange (EDI)

(*c*) Internet

(*d*) 1. Customer 2. Product 3. Supplier.

(*e*) 1. Business to Business (B2B)

2. Business to Customer (B2C)

3. Customer to Business (C2B)

4. Customer to Customer (C2C)

Q. 2. Give the reasons for increasing the popularity of E-com. Business Nowadays.

Ans. There are various attractions of e-commerce that has made it a popular Business tool. These are :

1. **Cost Effectiveness :** E- commerce transactions cost much lower to both the Order listing/ enumeration and customer service.

2. **Time Effectiveness :** When the search for a product is done then the features and specifications of similar items can also be viewed simultaneously. Thus transactions of a large number of items can be done easily which results in saving of time.
3. **More Information :** By providing information on a websites, more information can be dispensed to the customers, along with maintenance of proper record of their orders.
4. **Better Interaction :** Interaction with the customer is easier, faster and more descriptive. The customers can also confirm their orders and send related communication through E-mail.

Q. 3. Compare the advantage of E-commerce over traditional business.

Ans. Online purchasing and selling through e-business (wholesale and retail) Offers opportunities and advantages to the companies of all sizes engaged in different portfolios. Some of these advantages are enumerated below.

- Sampling of products such as books, recorded lecturers, and music cassettes is possible on Web for business promotion.
- Business houses selling their goods through catalogues, can reach additional global customers at lower cost.
- For items subjected to frequent changes e.g. shares and bonds etc., the changes in rates/quotations can be known instantly.
- Customer's knowledge can be enhanced on topics such as news about local event, market research, industrial report, software etc., which can be distributed easily over the Internet.
- A closer relationship can be developed amongst business sellers whose customer base is in on the Internet.
- For wholesalers, distributors, retailers etc., it is convenient and efficient to buy from Web sites.
- High quality, specialized products can be easily sold on Internet by the retailers.
- Organizations can conduct interviews through Internet to select suitable candidate(s) for their requirements.
- Group discussions can be conducted for corporate offices, industrial houses, business firms, universities, etc.

Q. 4. After setup business on the web, how a product can be purchased from any organization, give the answers in only few steps.

Ans. Few steps to set up Business on the Web.

1. The first requirement is an online store or commerce enabled website where goods or services can be described.
2. Online payment like payment through credit cards.
3. Packing and shipping after confirmation of order.
4. Customer service support like complaints returns.
5. Continuous promotion because without successful promotion your on-line business will fail.

Steps in product purchasing

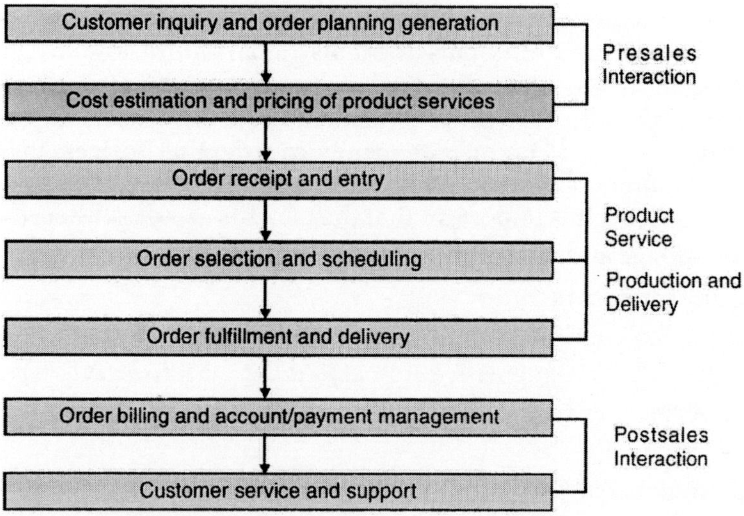

Steps in product purchasing E-Commerce Applications

Q. 5. How a business environment can be change by change in market, economical, societal and technological factors ?

Ans. The Driving Forces for E-commerce

Today's business environment is influenced greatly by market, economical, societal and technological factors creating a lot of competition. Also these factors are unpredictable in that they may change unexpectedly at any time. Such changes in the business world have been defined as *business pressure*. In order to succeed in this dynamic world companies must take action such as lowering costs and improving quality etc. At the same time they need to be innovative in customizing their products, creating new products and providing good customer service.

The various environmental business pressures on companies today can be grouped into three categories. Market, Societal and technological.

These are summarized in the table below :

Market and Economic pressure	– Strong competition
	– Global economy
	– Extremely low labor cost in some countries
	– Frequent changes in market demands
	– Increasing expectations of consumers
	– Awareness among consumers
Societal and Environmental Pressures	– Government regulations
	– Reductions in government subsidies
	– Rapid political changes
Technological Pressures	– Rapid technological obsolescence
	– New technologies
	– Information overload

Q. 6. Discuss the Benefits of E-Commerce.

Ans. Benefits of Electronic Commerce: Electronic commerce offers many potential benefits to organizations as well as individuals due to its global nature, low cost, interactive nature, variety of possibilities and resourcefulness and rapid growth of the supporting infrastructure, especially the web. These benefits have only started to materialize, and will increase significantly as E Commerce expands. In order that business invest their resources to engage in E-Commerce activities and consumers adopt the internet more and more for their shopping needs, it is important that the benefits exceeds the costs.

Q. 7. State the advantage and disadvantage of home banking.

Ans. Home banking services include :

- Account statement reporting
- Round the clock banking with automated teller machines (ATMs)
- Fund transfer
- Status of payment
- Tax return preparation
- Currency trading
- Debit or credit card management

These services do not yet have a very wide acceptance, even though they promise to cut the time spent drastically and are much more sophisticated than the traditional methods. The main reason for this is lack of technological awareness, costs involved for the common man and lack of trust is the new system. Hence for the success of consumer oriented e-commerce applications, consumer awareness and education is needed.

Q. 8. Differentiate between Mobile Commerce and E-Commerce.

Ans. Mobile commerce is concern with e-commerce applications via the media of wireless and mobile computing. A mobile commerce environment allows the users to access personal or business information and carry out all e-commerce activities while traveling away from home or workplace. The key features of mobile computing environment is that the user need not maintain a fixed position in the network.

Q. 9. Why has web security been given the top priority ?

Ans. In order to counter the various types of security threats/attacks to transmitted electronic data, security services to ensure reliable and true worthy transmission of digital business data and messages are needed. The primary security services are divided into five categories :

- Confidentiality
- Integrity
- Authentication
- Non-repudiation
- Authorization
- Availability

Q. 10. What are the legal requirements in E-commerce ?

Ans. Legal requirements in E-Commerce :

- Provisions of the agreement between parties.
- Course of dealing between the parties.
- Accepted trade practices in the applicable field of commerce.

- The civil code of the Jurisdiction or the common law as applicable.
- Pertinent statuary laws and regulations such as UCC and other legislative and administration enactments.

Q. 11. What is worldwide web ?

Ans. WWW basic concepts.

- URL-Uniform Resource Locater
- Protocol-http://
  ftp://
  Gopher://
  telnet://
- Host name
- Folder name
- File name
- Web Browsers
- IP Address
- Domain name

Q. 12. Why is the Electronic Job Market so popular especially among the high-tech profession ?

Ans. The Electronic Job Market so popular especially among the high-tech profession for some reasons to prevention of :

1. Destruction of information
2. Unauthorized availability of information
3. Authentication
4. Privacy/confidentiality
5. Integrity
6. Non-Reputability

Q. 13. Describe the E-Commerce infrastructure and Support services.

Ans. Network Infrastructure for E-commerce : The Network Infrastructure is the back bone of Electronic Commerce. A Secure, widely accessible, fully interconnected high speed International network that will guarantee the availability of bandwidth sufficient for the requirements of electronic commerce. For handling e-commerce transactions on the web various broadband technology options are available.

Q. 14. Discuss the role of E-Commerce in the light of business-to customer and business to-business relationships.

Ans. Based on the type of parties (business, consumers, etc.) involved, E-commerce may be of the following types:

1. **Business to Business (B2B) :** This type of electronic commerce includes the inter-organizational transactions and electronic market transactions between organizations or business.
2. **Business to Consumer (B2C) :** These are retailing transactions with individual shoppers. An online shopper buying from Amazon.com is **an Example of B2C e-commerce.**

Q. 15. List the different modes of Electronic Payment.

Ans. Different Modes of Electronic Payment : There are three types of payment systems have been established on the Internet :

1. Instant Paid or cash
2. Debit or Pre-paid
3. Credit or Post-paid.

Q. 16. Discuss various Encryption techniques used for web security.

Ans. In order to counter the various types of security threats/attacks to transmitted electronic data, security services to ensure reliable and true worthy transmission of digital business data and messages are needed.The primary security services are divided into five categories :

1. Confidentiality
2. Integrity
3. Authentication
4. Non-repudiation
5. Authorization
6. Availability

Q. 17. Discuss the benefits of Supply Chain Management.

Ans. The Benefits of Supply Chain Management : Supply Chain Management has three basic business benefits :

1. To get the right product, to right place, at least cost.
2. To keep inventory as low as possible.
3. To reduce the cycle times between customer orders and dispatch of finished products.

Q. 18. Discuss the various driving forces behind E-commerce.

Ans. The various driving forces behind E-Commerce.

1. Market and Economic Pressure.
2. Societal and environmental pressures
3. Technological pressures

Q. 19. Discuss the importance of Mobile Commerce.

Ans. The Importance of Mobile Commerce : The wireless is becoming more and more popular in E-Commerce Environment because of the following benefits:

1. Mobility
2. Ease of installation in difficult to wire areas
3. Reduced installation time
4. Increased reliability
5. Long term cost saving.

Q. 20. Define E-Commerce. Also list any five Commercial websites.

Ans. Electronic Commerce is an emerging concept that describes the process of buying and selling or exchanging of products, services and information via. Telecommunication and computer networks including the Internet. Five Commercial websites :

1. http://www.ebay.com/
2. http://www.baazee.com/
3. http://www.aol.com/
4. http://www.yahoo.com/
5. http://www.amazon.com/

Q. 21. What is Mobile Commerce ? Discuss its advantages.

Ans. Mobile Commerce : Mobile commerce is concern with E-commerce applications via the media of wireless and mobile computing. A mobile commerce environment allows the users to access personal or business information and carry out all e-commerce activities while traveling away from home or workplace.

Benefits :

1. Mobility
2. Ease of installation in difficult to wire areas
3. Reduced installation time
4. Increased reliability
5. Long term cost saving.

Q. 22. State the difference between inter and intra Organizational E-Commerce.

Ans. Difference between Inter and Intra Organizational E-Commerce : An inter organization involves information flow among two or more Organizations. Major objective is efficient and effective transaction processing like Transmitting orders, bills and payments using electronic data exchange. Intra organizational E-commerce means intra business. It includes all internal organizational activities generally performed on intranets. It includes exchange of goods, services or information.

Q. 23. Discuss the role of Firewall.

Ans. The Role of Firewall : A method for implementing security policies designed to keep a network secure from intruders. It can be a single router that filters out unwanted packets or may comprise a combination of routers and servers each performing some type of firewall processing. Firewalls are widely used to give users secure access to the internet as well as to separate a company's public web server from its internal network.

Q. 24. Discuss the Security Attacks, Services of Web Security.

Ans. Security Attacks, Services of Web Security : Different organizations and individual may face different types of threats to the security of their information on the Internet. Some of the common risks/threats are :

- Data may be stolen, corrupted misused, altered.
- Attacks on hardware may occur that may render systems unable to operate properly.
- Programs may be altered to cause systems to perform incorrectly or even to crash.

Q. 25. What are the different EDI applications in business ?

Ans. The different EDI applications in business :

- In Financial Matter
- In Electronic Fund Transfer (EFT)
- In Automatic Clearing House (ACH) transfer.

Q. 26. Explain, Clearly, the concept of Supply chain management giving suitable example.

Ans. Supply Chain Management : A supply chain management is a collection of interdependent steps that, when followed, accomplished a certain objectives such as meeting customer requirements. The supply chain is gaining prominence as manufacturers control less of the speed at which products are manufactured and distributed.

Q. 27. Discuss, briefly, the importance of building the customer relationship. Support you answer by giving suitable examples.

Ans. There are several reasons :

- Choice
- Convenience
- Better Consumer Information.

Q. 28. What is WAP technology ?

Ans. WAP Technology (Wireless Application Protocol) : WAP is an evolving standard that accelerates mobile commerce market with a WAP capable mobile device. Someone can do Business transaction like shopping, banking operations, reservations on Internet WAP is a result of specific work done by WAP forum. WAP is a standard for providing cellular phones, pagers and other handheld devices with secure access to e-mail and text based web pages.

Q. 29. Discuss the various mobile information devices.

Ans. Various mobile information devices :

- Portable Computers
- Hybrid Pen Computers
- Personal Digital Assistance (PDA)
- Digital Assistance
- Personal Communicators
- Palmtops

Q. 30. Define Encryption. Discuss, in detail, the Public Key encryption.

Ans. Encryption : Encryption is the process of transforming information. So it is unintelligible to anyone but the intended recipient.

Public Key Encryption : A more powerful form of cryptography involves the use of public keys. Public key techniques involve a pair of keys—private key and public key associated with each user. The public key is made available to anyone who wants to send a encrypted message to the holder of the private key. The only way to decrypt the message is with the private key of the pair.

Q. 31. Discuss the security issues involved with different electronic payment modes.

Ans. The security issues involved with different electronic payment modes :

- Physical Security holes
- Software Security Holes
- Inconsistent uses holes
- Trust Based Security
- Security through Obscurity (STO)
- Password Schemes

Q. 32. What is Home Banking ?

Ans. Home Banking : The concept of Home Banking is akig from home itself. Customers use a numeric password on a push Button telephone to access Banking services. There was one drawback of this banking is o visual verification possible, which is important to customer.

Q. 33. Discuss the advantage and disadvantage of Home Banking.

Ans. Advantage and disadvantage of Home Banking :

1. View account balance
2. Access account history
3. Transfer funds between accounts
4. Schedule future transfer
5. Pay ills electronically
6. Schedule automatic recurring bill payment

Q. 34. Discuss the importance of client server network model. How Security be achieved with this model ?

Ans. Client Server Network Model : The client/server architecture of the Internet consist of many nodes which are either servers or clients, providing either a set of services or requesting a set of services.

Clients are mostly the computers from which the users communicate to other nodes which may be other clients or servers. The servers are the centralized service providers that offer services for the clients.

The primary security services are divided into five categories :

1. Confidentiality
2. Integrity
3. Authentication
4. Non-repudiation
5. Authorization
6. Availability

Q. 35. Describe the Network Infrastructure for E-Commerce.

Ans. The Network Infrastructure for E-Commerce : The Network Infrastructure is the back bone of Electronic Commerce. A secure, widely accessible, fully interconnected high speed network that will guarantee the availability of bandwidth Sufficient for the requirements of electronic commerce. For handling e-commerce transactions on the web various broadband technology options are available.

Q. 36. What do you mean by Customer relationship management (CRM) ?

Ans. Customer Relationship Management (CRM) : Customer relationship management (CRM) is concerned with "Who the customers are," "What the do", What they want" and "What are they are willing to pay for" ?

Some facts about customer relationship are :

- It consists six times more to sell to a new customer than to sell to an existing one.
- A dissatisfied customer will tell many others about his experiences.

Q. 37. What are the main types of e-commerce ? Explain.

Ans. The Main Types of E-commerce : Depending upon the type of transactions, the electronic commerce applications are divided into two main categories :

1. Buying and selling goods and services, usually referred to as electronic market.
2. Facilitating inter and intra organization flow of information, communication and collaboration. These are referred as inter organizational system.

Q. 38. Describe the Categories of E-Commerce based on Transacting Parties.

Ans. Categories of E-Commerce Based on Transacting Parties : Based on the type of parties (business, consumers, etc.) involved, E-Commerce may be of the following types :

- Business-to-Business (B2B)
- Business-to-Consumer (B2C)
- Consumer-to-Consumer (C2C)
- Consumer-to-Business (C2B)
- Intra business (Organizational)

Q. 39. What are the major parts of I-way infrastructure ?

Ans. The Major Parts of I-way Infrastructure : The I-way or information super high way consists of components which can be broadly categorized as :

- Network Access Equipment
- Access Roads or Media
- Global Information Distribution Networks
- Consumer Access Equipment

Q. 40. Describe the types of Firewalls.

Ans. The Types of Firewalls : There are several types of firewalls, which offer various levels of security. The most widely used method of fire walling is to place a computer or a router b/w the n/w and the Internet, which will then control and monitor all traffic b/w the outside world and the local n/w.

- Simple Traffic Logging System
- IP pocket Screening Routers (Pocket filtering gateway)
- Incoming packet protocol
- Hardened Firewall Host
- Proxy Application Gateways

Q. 41. What is WAP(Wireless Application Protocol) ?

Ans. WAP (Wireless Application Protocol) : WAP is an evolving standard that accelerates mobile commerce markets with a WAP capable mobile device. Someone can do business transaction like shopping, banking operation, reservation on Internet WAP is a result of specific work done by WAP forum. WAP is a standard for providing cellular phones, pagers and other hand held devices with secure access to e-mail and text based web pages.

Q. 42. Describe the features of Wireless Application Protocol.

Ans. The Features of Wireless Application Protocol :

- Wireless application protocol (WAP) is an open communication standard that is recognized globally.
- Wireless application protocol allows for communication between a mobile handset and the Internet.
- The device includes mobile phones, two-way radios, smart phones and communicators.
- WAP operates in much of same fashion as the web does.

Q. 43. Describe the various types of channels used in network Infrastructure for E-Commerce.

Ans. The Various Types of Channels Used :

- Twisted Pair (Shielded/Unshielded)
- Coaxial Cable
  - (*a*) Base band Coaxial Cable
  - (*b*) Broadband Coaxial Cable
- Fiber optic cable
  - (*a*) Single Mode Fiber optic cable
  - (*b*) Multiple mode Fiber optic cable
- Radio Transmission
- Microwave
- Infrared

Q. 44. Explain threats to security.

Ans. Threats to security : Threats to security consists of unauthorized modification of server data, modification of incoming data packets and compromise of a server system by exploiting bugs in the server software.

- Hackers have potential access to a large no of systems. As a result, computers with security holes are particularly vulnerable.
- Hackers can use electronic eves dropping to trap user names and unencrypted passwords sent over the network.

Q. 45. Explain the difference between a packet-filtering firewall and an application-level proxy server.

Ans. The difference between a packet-filtering firewall and an application-level proxy server : Packet filter blocks traffic based on a specific WEB address (IP address) or type of application (e-mail, ftp, Web, etc.), which is specified by port number. Also Known as a "screening router." Proxy server serves as a relay between two networks, breaking the connection between the two. Also typically caches Web Pages.

Q. 46. Explain Virtual Private Networks.

Ans. Virtual Private Networks : Since the extranet transmission is over the Internet, there is little privacy or transmission security. Therefore, when using an Extranet, it is necessary to improve the security of the connected Intranets and the communication channel. This is achieved through techniques of cryptography and firewalls. Such Internet channels connecting two intranets and secured through cryptography and firewall techniques are termed as Virtual Private Networks (VPNs).

Q. 47. Explain smart card as Internet payment method.

Ans. Smart Card : Smart Cards are prepaid credit cards that may include an embedded microprocessor. These microprocessor card can add, delete and otherwise manipulate information on the card while a memory chip card can only undertake predefined operation. Smart Card, unlike magnetic stripes cards can carry all necessary functions and information on the card. There fore they do not require access to remote databases at the time of the transaction.

Q. 48. Describe the components of WAP.

Ans. The Components of WAP : The WAP protocol stack consists of the following components :

(*a*) WML
(*b*) WML script
(*c*) WBMP
(*d*) WBXML
(*e*) WBP/B
(*f*) WSP
(*g*) WTP
(*h*) WTLS
(*i*) WDP

Q. 49. Define SCM (Supply Chain Management) and also describe its components.

Ans. SCM (Supply Chain Management) : SCM (Supply Chain Management) is the co-ordination of material, information and financial flows among all the participating enterprises in a supply chain. Its components :

- Supplier Management
- Inventory Management
- Distribution Management

- Channel Management
- Payment Management
- Sales Force Management

Q. 50. Give the layered structure of EDI. Also describe the benefits and Components of EDI.

Ans. **Benefits of EDI :** The benefits from EDI implementation can be classified into direct benefits and long-term strategic benefits.

Direct Benefits :

- Improve customer service is enabled.
- Business documents move faster.
- Information is managed more efficiently.

Strategic Benefits :

- The competitive edge is maintained and enhanced.
- Reduction in product costs is achieved.
- More accurate forecasting and business planning is achieved due to information availability at the right place ad right time.

Q. 51. Explain the Global Information Distribution networks.

Ans. The global information distribution networks consists of the Infrastructure criss-crossing the countries and continents. They include the log distance telephone lines, satellite networks, and the Internet.

- Long Distance Networks
- Satellite Networks

Q. 52. What is the Demand of Broad Band Technology in Ecommerce ?

Ans. **The Demand of Broad Band Technology in Ecommerce :** As the Internet market continues to explode, demand for greater Bandwidth and faster connection speeds have led to several Technological approaches developed to provide broadband Access to all consumers, the demand coming primarily from. Three sources :

- Small and mid size businesses and small offices/home offices.
- Consumers, and multiple-tenant units (MTUs).
- In building fiber builders and universities that are installing high-speed wireless connections throughout their campuses.

Q. 53. Explain the various components of the I-way Infrastructure.

Ans. **Components of the I-way Infrastructure :** The I-way or information super high way consists of various components which can be broadly categorized as :

- Network Access Equipment
- Access Roads or Media
- Global Information Distribution Networks
- Consumer Access Equipment

Q. 54. Define and differential TDMA (Time Division Multiple Access), FDMA (Frequency Division Multiple Access) And CDMA (Code Division Multiple Access).

Ans. **FDMA (Frequency Division Multiple Access) :** In a FDMA system, different frequency bandwidths are assigned to different users. The FDMA system has been widely adopted in existing analog cellular systems.

TDMA (Time Division Multiple Access) : TDMA splits a single carrier wave into time slots and distributes the slots among multiple users.

CDMA (Code Division Multiple Access) : CDMA is a wideband spread spectrum technology that spreads multiple conversations across a wide segment of the cellular broadcast spectrum. CDMA used by digital cellular services, uses digital encoding to multiplex several channels on to one line.

Q. 55. What are digital Signatures ?

Ans. Digital Signatures : In situations where there is not complete trust between the sender and the receiver, something more than verifying message integrity is needed. The most attractive solution of this problem is the digital signature, i.e. digital signature are designed to bind the message originator with the exact contents of the message. A digital signature may be defined as an encrypted message digest.

Note. After the Question answer session, practice can be done through assignment.

❑❑❑

Assignment No. 1

Q. 1. What is Electronic Commerce ? Describe five potential benefits of Electronic Commerce for business.

Q. 2. Describe the architectural framework of Electronic Commerce. What are the Applications of this ?

Q. 3. Describe Intra Organizational Commerce and its applications in the area of Business to Employee services.

Q. 4. What is the Global Information Distribution Network ? What are the issues of Public Policy Shaping the I-way ?

Q. 5. Explain the various issues involved in Electronic Payment System. Also give the various types of EPS.

Q. 6. What do you understand by protocol and briefly describe the role of TCP/IP in internet ?

Q. 7. What is the basic Network Infrastructure for E-Commerce ?

Q. 8. Describe the Market forces influencing the I-way.

Q. 9. What is Global Information Distribution Network ? What are the issues of Public Policy Shaping the I-way ?

Q. 10. Write the short notes of any two :

(*a*) Online Banking

(*b*) Digital Signature

(*c*) Legal requirement in E-commerce

Assignment No. 2

Q. 1. Give the layered structure of EDI and compare EDI versus e-mail. Conceptualize EDI and compare EDI versus e-mail. Discuss the Legal issues related with e-commerce.

Q. 2. What do you mean by Mobile Commerce ? Which type of Protocols are necessary for wireless application ?

Q. 3. What do you mean by Encryption ? How Encryption is useful in Virtual Private Network (VPN) ?

Q. 4. Discuss the importance of client server network model. How can Security be achieved with this model ?

Q. 5. (*a*) What is WAP technology ? Discuss the various mobile Information devices.

(*b*) Discuss the benefits of Supply Chain Management.

Q. 6. List the different modes of Electronic Payment. Explain in detail the digital token based electronic payment system.

Q. 7. What do you understand by customer relationship management ? What importance has customer relationship mgmt got in E-Commerce ? How will you implement customer relationship mgmt in your Organization ?

Q. 8. Describe Intra organizational Commerce and its applications in the area of Business to Employee services.

Q. 9. Describe the architectural framework of electronic commerce. What are the Application of this ?

Q. 10. Short notes :

(*a*) Legal requirement in E-Commerce

(*b*) Firewall and network security

(*c*) World Wide Web and Security

Assignment No. 3

Q. 1. Discuss the importance of Mobile Commerce. Explain various Mobile information devices in the light of Mobile Commerce.

Q. 2. Describe the Layered architecture of EDI in E-Commerce. What are the benefits of EDI over traditional business, also describe the Application of EDI in business.

Q. 3. What do you mean by Encryption ? Discuss various Encryption Techniques used for web security.

Q. 4. State the difference between Mobile Commerce and E-Commerce. Explain various applications of Mobile Commerce in the current Scenario.

Q. 5. (*a*) Explain, clearly, the concept of supply chain management giving Suitable example.

(*b*) Explain the difference between a packet-filtering firewall and an Application-level proxy server.

Q. 6. Discuss the various driving forces behind e-commerce. Also explain the industry framework and the practical problem in implementing the e-commerce.

Q. 7. Critically comment on E-commerce as a new and effective way for conducting business product and providing better customer services on the Internet.

Q. 8. What is the likely impact of electronic commerce on economic structures like manufacturers, dealers and consumers ?

Q. 9. What is system reliability assurance also describe the characteristics of reliable Information system ?

Q. 10. Short notes :

(*i*) Firewall and network security

(*ii*) Client-Server security threats

(*iii*) Digital Signature

Assignment No. 4

Q. 1. (*a*) Discuss the client-server architecture of E-commerce, is client-server architecture a necessity for E-commerce.

(*b*) Write short note on :

- Fiber optics
- Satellites

Q. 2. (*a*) Discuss the mechanism of Video on demand system/server.

(*b*) Write short notes on :

- Virus
- Trojan Horse
- Worm

Q. 3. (*a*) What are the various components of the I-way, discuss it along with a proper diagram.

(*b*) Differentiate between E-Commerce and M-Commerce.

Q. 4. (*a*) Define and explain:

- Just-in-Time Manufacturing
- Quick Response Retailing

(*b*) What do we mean by a firewall, where it is located, what are the various types of firewalls employed in practice, what are the various rules governing the firewalls.

Q. 5. (*a*) Differentiate between Set-top Boxes and a PC.

(*b*) Define the various statistics of a website in points.

(*c*) Discuss the HTTP mechanism to connect to the Internet, can we connect the internet without using HTTP connection.

Q. 6. (*a*) What do we mean by mobile code and message overloading.

(*b*) Discuss the Electronic Commerce Framework, what is specified by the two Poles in the framework.

Q. 7. (*a*) Discuss the role of E-commerce in light of business to customer and business.

(*b*) Describe the E-commerce infrastructure and support services.

Q. 8. What is Electronic Commerce ? Describe five potential benefits of electronic Commerce for business.

Q. 9. What is the Customer-Oriented value chain ? Describe three forces lead to the www's emerges as the single most dynamic force in information technology.

Q. 10. Short Notes :

(I) EDI

(II) Distributed Network

(III) Authentication

(IV) Quick response retailing

(V) Client Server Network

(VI) Digital Signature

(VII) Video on Demand

Assignment No. 5

Q. 1. (*i*) What is WAP; discuss it with reference to mobility.

(*ii*) What do we mean by TDMA, what technology is used in TDMA. List the various benefits of TDMA over CDMA.

Q 2. *(i)* Write a short note on EDI, what are the various benefits of EDI. Can EDI be done without Electronic Commerce infrastructure ?

(ii) Discuss the layered architecture of EDI along with a proper diagram. What are X.12 and EDIFACT in EDI ?

Q. 3. *(i)* What are the two broad areas of security concerns in E-Commerce ?

(ii) What are the various protection mechanisms employed for security.

(iii) Discuss Firewalls and Network Security.

(iv) What is health care EDI ?

Q. 4. *(i)* Discuss the Electronic Commerce framework (various building blocks in the infrastructure), show it using a proper diagram.

(ii) What are the various components of the I-Way, discuss in short.

(iii) What do we mean by e-cash, what are the various properties of e-cash, what are the various problems associated with e-cash.

Q. 5. *(i)* What do we mean by CDMA, CDMA is based on wireless or wire based technology illustrate it, why CDMA is called wide band and spred spectrum technology, discuss it.

(ii) Discuss cellular communication; what is the "Hand Off" situation in cellular communication, what is the solution to it.

Q. 6. *(i)* Write a short note on any one of the following:

(ii) Digital token based Electronic payment systems.

(iii) Smart cards based Electronic payment systems.

(iv) Differentiate between EDI and E-mail.

Q. 7. *(i)* What is the difference between public key encryption and private key encryption ?

(ii) Discuss the importance of E-commerce.

(iii) State the advantage and disadvantage of home banking.

(iv) What are the legal requirements in E-commerce ?

Q. 8. *(i)* What is online banking ? State the merits and demerits of online banking.

(ii) State the difference between inter and intra organizational E-commerce.

(iii) What are the two pillars of e-commerce business ? Describe in detail

Q. 9. *(i)* Discuss the WAP application architecture i.e. how the WAP device accesses the application server.

(ii) Explain smart card, as Internet payment method. What are the components of smart card chip ?

Q. 10. *(i)* Define E-Governance, what do we mean by G-to-C and G-to-G E-commerce methods.

(ii) What is the difference between online and offline transaction ? How does ecommerce support then online transaction ?

❑❑❑

EXAMINATION PAPERS

B.TECH.

FIFTH SEMESTER EXAMINATION, 2006-07

E-COMMERCE

Time : 3 Hours *Total Marks : 100*

Note :

(i) *Attempt ALL questions. All parts of a question should be attemped at single contagious place.*

(ii) *All questions carry equal marks.*

(iii) *In case of numerical problems assume data wherever not provided.*

(iv) *Be precise in your answer.*

1. Attempt *any four* parts of the following : **(5 × 4 = 20)**

(*a*) What is intra organisational E-commerce and what are its potential benefits ?

(*b*) Which type of transactions come under the category of C2B e-commerce ?

(*c*) "E-commerce is the buying and selling of information, products and services via computer networks." Does this definition include all aspects that you would consider as part of E-commerce ? Of why or why not ?

(*d*) What is supplier-centric B2B e-commerce.

(*e*) Discuss the role of I-way in development of E-commerce.

(*f*) What are Intranet and Extranet ? What are their potential advantages ?

2. Attempt *any four* parts of the following : **(5 × 4 = 20)**

(*a*) Why is the internet vulnerable to hackers ? Describe various attack types web-serve is prove to ?

(*b*) What are threats posed by a DNS spoofing attack ?

(*c*) Compare the stateful inspection firewall with the application level proxy firewall.

(*d*) Discuss the main features of WAP technology.

(*e*) How does AMPS differ from GSM ?

(*f*) Compare and contrast the concepts of universal personal telecommunications and universal communications access.

3. Attempt *any two* parts of the following : **(10 × 2 = 20)**

(*a*) Compare Secure Socket Layer (SSL) and Secure Electronic Transaction (SET) protocols. Why Payment Cards (debit, credit and charge cards) are much more popular for online payments then e-cash and e-wallets.

(*b*) Propose a Security Model for an E-Banking Application.

(*c*) What is Public Key Infrastructure ? Describe the role of the certification authority and registration authority.

4. Attempt *any two* parts of the following : **(10 × 2 = 20)**
   (*a*) Describe the functioning of a fully integrated EDI system. What are its benefits over a Non-EDI based system ?
   (*b*) What are the major components of supply-chain management ? How E-commerce can be helpful in each of them ?
   (*c*) Discuss various types of E-payment systems and their comparative advantages and disadvantages.

5. Write short notes on *any four* : **(5 × 4 = 20)**
   (*a*) Emerging Financial Instruments.
   (*b*) Virtual Private Networks (VPN).
   (*c*) CDMA technology.
   (*d*) Broad Band Telecommunications.
   (*e*) Role of multimedia in development of E-commerce.
   (*f*) Digital Signature.

B.TECH.

FIFTH SEMESTER EXAMINATION, 2007-08

E-COMMERCE

Time : 3 Hours *Total Marks : 100*

Note : *Attempt ALL questions. All parts of a question should be written at single contiguous place.*

1. Attempt *any four* parts of the following : **(5 × 4 = 20)**
   (*a*) What are the benefits of E-commerce ? Also write down the limitations of E-commerce.
   (*b*) What are internet, intranet and extranet ? List advantages a company can derive from using each of the three.
   (*c*) Discuss the role of communication technologies in development of e-commerce ? What is the last mile in communication technology term ?
   (*d*) How e-commerce influences the elements of a market ?
   (*e*) What is B2B e-commerce ? Who are the players involved in it ?
   (*f*) What is the role of certification authority in E-commerce framework ?

2. Attempt *any four* parts of the following : **(5 × 4 = 20)**
   (*a*) Characterize the three generations of mobile telephony.
   (*b*) What are some ways of achieving frequency reuse in wireless communication ?
   (*c*) Describe what a denial of service attack is and how it affects e-commerce. How it can be prevented ?
   (*d*) What are major threats posed by a sniffing attack ?
   (*e*) Discuss the advantages of using a proxy server.
   (*f*) What is meant by DMZ ? What are its advantages and disadvantages ?

3. Attempt *any two* parts of the following : (10 × 2 = 20)
   (*a*) What are major issues in e-mail security ? What are various schemes that are used to ensure the privacy of e-mails ?
   (*b*) What is a digital certificate ? Describe the commonly used standards for digital certificate.
   (*c*) Universal access, regulation and information privacy are three major issues that are shaping the I-way. Discuss the positive and negative aspects of these issues. What do you see as the future of the Internet in these areas ?
4. Attempt *any two* parts of the following : (10 × 2 = 20)
   (*a*) Discuss the role of EDI in B2B e-commerce. How can Internet-EDI affect the balance of power in businesses ?
   (*b*) Extranets are designed to allow companies to share data among themselves. How can a firm share data over their Extranet and the Internet and still restrict access to hackers ?
   (*c*) Explain how a hybrid approach using both DES symmetric keys and RSA public-private key pairs works ?
5. Write short notes on *any four* : (5 × 5 = 20)
   (*a*) Digital signautre.
   (*b*) Transaction security in E-commerce.
   (*c*) Market forces behind I-way.
   (*d*) WAP technology.
   (*e*) E-Wallets.
   (*f*) E-Advertising.

❑❑❑

INDEX

A

A Sample SET Session 97
Access Media 45
Advantages of Broadband 46
An SET Implementation 99
Application Gateways or Proxy Gateways 76
Application Layer 64
Application Level Protocols 28
Applications of M-Commerce 56
Applications of Mobile Computing 61
Authentication 89
Automated Clearing House (ACH) Transfers 122
Awareness, Education and Skills 142

B

Bank Checks 122
Banks 18
Basic Component of SCM 123
Basic VPN Requirements 91
Bearer Interfaces 65
Benefits of Electronic Checks 109
Broadband over Powerline (BPL) 48
Building Blocks of EDI Systems : Layered Architecture 38
Business to Business E-Mail 16
Business to Customer E-Mail 16
Business-to-Customer Electronic Commerce 14
Business-to-Employee (B2E) Services 15

C

Cable Modem 47
CDPD Network Reliability 60
CDPD Technology (Cellular Digital Packet Data) 59
CGI Scripts, CGI Programs or CGI Applications 34
Changing Consumer Needs 111
Changing Dynamics in the Banking Industry 111
Clarification and Revision of Laws and Regulations 143
Client server Network Security 72
Client Server Technology 70
Client/Server Model of Computing 23
Code of Social Ethics 131
Commercial Opportunities 9
Common Gateway Interface 34
Common Payment Gateways 101
Common Uses of VPNs 90
Component of E-Commerce 18
Components of the I-Way 45
Confidentiality Breaches 73
Connect to a Telnet Address 29
Connecting Computers Over an Intranet 91
Connecting Networks Over the Internet 90
Consumers 18
— Behaviour 56
— Issues 133
Consumer-to-Consumer Electronic Commerce 15
Coordinating Public and Private Sector Electronic Commerce Activities 142
Cost Reduction 112
Credit Card Fraud 80
Customer Relations Management (CRM) 126
Customer to Customer E-Mail 16
Customer's Passwords 68
Customer-to-Business Electronic Commerce 15
CyberCash/CyberCoin 107

D

Data Communications 58
Database Vulnerabilities 67
Debit Cards Based EPS 109
Defining EDI 118
Defining New Rules and Regulations for an Electronic Environment 142
Defining the Policy and Regulation for Electronic Commerce 140
Definition of Ethics in E-Commerce 129
Denial-of-Service 73
Dependency on location and Time 55
Designing of a Well-Functioning Commercial Website 144
Destructive Behavior 73
Developing E-Commerce Enabled Application 145
Device Constraints 55
DigiCash (or E-cash) 106
Digital Certificate 84
Digital Signature 85
Digital Subscriber Line (DSL) 47
Digital Time-Stamping Service 86
Digital Token-based Electronic Payment Systems 103
Domain Name System 23
Domain Names 22
Driving Force for M-Commerce 54

E

E-Commerce Laws in India 138
E-Commerce Laws 131
E-Commerce Security Issue 67
E-Commerce: Government's Policies and Agenda 140
EDI (Electronic Data Interchange) 16
EDI in Action 120
EDI Layered Architecture 119
EDI Over Wan/Internet/Extranet 16
EDI Versus E-mail 119
EDIFACT 122
EFT Services 110
Electronic Checks Action 109
Electronic Checks 108
Electronic Commerce and Banking 111
Electronic Commerce and Online Publishing 114
Electronic Commerce and Retailing 115
Electronic Commerce Definition 2
Electronic Commerce Framework 12
Electronic Data Interchange 118
Electronic Data Interchange 38
Electronic Fund Transfer 110
Electronic Funds Transfer (EFT) 17
Elements of an Encryption System 82
Elements of E-Commerce 5
Emerging Client Server Security Threats 72
Encryption 82
Encryption and Decryption Laws 137
Enforcement of Laws 143
Entertainment 10
Executing Commands Illegally 73
Existing Cellular Network Architecture 57
Extended Supply Chain 125

F

Fiber 48
File Transfer Protocol 29
Financial EDI 121
Financial Institutions 18
Finding a Provider of Online Transactions 149
Finding Shopping Cart Software 150
Firewall Protection 74
Firewalls and Network Security 74
Flexibility in Public/Private Sector Relationships 142
Forms of Agreements 135
Frame Relay Devices 51
Frame Relay Introduction 50
Frame Relay Standardization 51

G

Getting a Merchant Bank Account 147
Getting an Internet Connection 146
Global Information Distribution Network 45
Government 18

H

History of Electronic Commerce 1
Home Banking Implementation Approaches 112
How Payment Gateways Work 101

HTTP 28
Hybrid Systems 77
Hypertext Markup Language (HTML) 35

I

Impact of Globalization on the Supply Chain 125
Implications of Various Factors 4
Importance of Firewall 74
Improving Social Ethical Climate 130
Institutions of E-Commerce 18
Intellectual Property Issues 133
Interbank Electronic Funds Transfer (EFT) 122
Interconnection Agreements 136
Intermediate Systems (IS) 60
International Chamber of Commerce (ICC) 138
Internet and Intranet Based E-Commerce Issues 42
Internet History 20
Internet Infrastructure and Routing 23
Internet Transactions and Security 78
Internet 20
Intra/Inter Organizational E-Commerce 15
Intra-Organizational Mail 16
Introduction to
— ATM 49
— Broadband 46
— ISDN 49
— I-Way 44
— Mobile Commerce 53
— VPN 89
— WAP (Wireless Application Protocol) 62
IP Addresses 22
Issue of
— Cyber Jurisdiction 134
— Copyrights, Trademarks and Trade Names 132
— CRM 127

L

Lack of Standards 55
Legal Disputes on the Internet 133
Legal Status of Documents Signed with Digital Signatures 87
Legislation Dilemma 43
Limitations of Firewall 75
Limiting Factors 55

M

Major Threats and Challenges to Social Ethics 130
Management Issues in Online Banking 113
Market and Economic Pressure 3
M-Commerce Versus E-Commerce 54
Mechanics of SET 98
Merchant Account versus Bank Accounts 147
Merchants 18
Mobile Computing 56
Mobile Customer and Applications 55
Mobile Data Base Station (MDBS) 60
Mobile Data Intermediate Systems (MDIS) 60
Mobile End User Interface 60
Mobile Information Devices 65

N

Need of Digital Certificate 85
Need of Online Publishing 114
Network Access Equipment 45
Network for E-Commerce 44

O

Objectives of SET 98
Obtaining a Digital Certificate 148
Online Credit Card Based System 104
Online Publishing Approach 115
Online Shopping 116
Online Terms, Conditions, Policies and Laws 43
Opportunities of E-Commerce 9
Overview of the Client Server Technology 71

P

Packet Filtering 76
Payment Gateway 100
Payments Agreements 137
PayPal 108
People Aspect of CRM 126
Personalization 55
Policy Recommendations for E-Commerce Growth 141
Practical Requirements for SET 100
Privacy Issue in E-Commerce 131
Privacy 42

Problems with Traditional
Payment Methods 97
Process and Technology
Aspects to CRM 127
Process of E-Commerce 19
Promoting Productive Industry/Government
Relations 142
Proper Action Plans to Reflect the Pace of
Technical Change 142
Protection of Intellectual Property 143
Public-Key Encryption 83

R

Reforming Regulatory Practices 142
Registering a Domain Name 148
Regulation of the Information Infrastructure 141
Remote User Access Over the Internet 90
Requirement of I-Way 44
Roadblocks to Installing Supply Chain
Software 124
Role of Governments to Promote
Electronic Commerce 140
Role of Web Server on Internet 33

S

Satellite 48
Secret-Key Encryption 82
Secure Electronic
Transaction (SET) Protocol 97
Secure Information Exchange 141
Secure Socket Layer (SSL) 87
Secure Socket Layer Protocol 31
Security Mechanisms are
Available Through VPN 93
Security Provision in Model Agreements 137
Security 42
Session Layer 64
Shopping Cart Software 150
Shopping Cart Vulnerabilities 68
Signing Digital Certificates 149
Smart Cards 108
Social Issues In E-Commerce 129
Social Services 10
Societal and Environmental Pressures 3
Some Emerging Technologies that
will Affect the Supply Chain 126
Standardization 141
Standardization and EDI 122
Strength of E-Commerce 6
Suggestions for EFT 110
Suppliers 18
Supply Chain
— Collaboration 125
— Management Software 123
— Management 123
Supporting Opportunities for the
Growth of E-Commerce 141
SWOT (Strength, Weakness,
Opportunity, Trends) Analysis 5

T

Tangible Benefit to EDI 121
Taxation Issues 133
TCP/IP Protocols 27
Technological Pressures 3
Telnet Protocol 29
The 'Physical' Infrastructure 142
The Designer's Liability 132
The Driving Forces for E-Commerce 2
The Issue of Liability 132
The Mechanism of Electronic Commerce 13
The SMTP Protocol 31
The User's Liability 132
The Web and E-Commerce 26
The Web Server Computer—The Hardware 33
The Web Server Program—The Software 33
Trading Partner Agreements 136
Transaction, Security, and
Transport Layers 64
Trends In E-Commerce 10
Tunneling Basics 92
Types of
— Broadband 47
— Credit Card Payments 104
— Electronic Commerce 14
— Electronic Payment System 102
— Financial EDI 121
— Firewalls 76
— Identity Theft 79
— Web Security Risks 69

U

Unauthorized Access 73
Uniform Resource Locator (URL) 21
United Nations on E-Commerce 138
Use of
— Digital Certificates 85
— Digital Signature for Authentication 86
— E-Mail in E-Commerce 15
— ISDN 49
— SSL 88
— Telnet 29
Using FTP to Download (Transfer from Another Machine to Yours) 30
Using FTP to Upload (Transfer from Your Machine to Another) 30
Using the Digital Currency 103

U

Validity of Digital Signatures 86
Value-Added Network (VAN) Agreements 136
Virtual PIN 106
Voice is Still the Killer Application 55
VPN Management Issues 93

W

WAP Architecture Overview 63
WAP Forum 63
WAP Protocol Stack 64
Weaknesses of E-Commerce 7
Web Architecture 25
Web Browser 32
Web Hosting 148
Web Linking and Domain Name Disputes 134
Web Security Precautions 70
Web Security 69
Web Servers 33
Why ISDN to Access the Internet 50
Wireless 48
Working of Broadband 46
World Wide Web Security 78
World Wide Web 25

❑❑❑

Win Prizes

ATTENTION READERS

Win Prizes

We request for your frank assessment regarding various aspects of the book as given below :

CONCEPTS OF

E-COMMERCE

By : ADESH K PANDEY

Please Fill up the Sheet in Neat and Clean Writing

(i) What topics of your syllabus, important for your examination point of view, are not covered in this book ?

....

....

....

....

(ii) What are the chapters wherein the treatment of the subject matter is not systematic or organised or updated ?

....

....

....

....

(iii) Have you come across misprints/mistakes/factual inaccuracies in the book ? Please specify the chapters and the page numbers.

....

....

....

....

(iv) What other books on the same subject have you found/heard better than the present book ? Please specify in terms of price and quality.

....

....

....

....

(v) Further suggestions and comments for the improvement of this book.

....

....

....

....

(PS : If need be please attach a separate sheet to write your views)

OTHER DETAILS

(i) Who recommended you this book ? (Please tick (✓) in the box near the option relevant to you.

☐ Teacher ☐ Friends ☐ Bookseller

....

....

....

....

(ii) Name of the recommending teacher, his designation and address:

....

....

....

....

(iii) Name and address of the bookseller you purchased the book from :

....

....

....

....

(iv) Name and address of your institution (Please mention the University or Board as the case may be)

....

....

....

....

(v) Your name and complete address :

....

....

....

....

(vi) Course you are enrolled in :

....

....

....

....

The best assessment will be awarded each month : The award will be in the form of our publications as decided by the Editorial Board.

Please mail the filled up coupon at the vearliest to :

S.K. KATARIA & SONS®

4885/109, Prakash Mahal, Dr. Subhash Bhargav Lane,
Opposite Delhi Medical Association, Daryaganj, New Delhi-110002
Phone: +91-11-43551243, Mobile : +91-9871775858, 7678425209
email : katariabook@gmail.com, katariabooks@yahoo.com